Appli~~ed Anatomy~~ for Anaesthesia and Intensive Care

Applied Anatomy
for Anaesthesia and
Intensive Care

Andy Georgiou FRCA DICM EDIC FFICM
Consultant in Anaesthesia and Intensive Care Medicine
Royal United Hospital Bath NHS Trust

Chris Thompson FRCA EDRA
Consultant in Anaesthesia
North Bristol NHS Trust

James Nickells FRCA
Consultant in Anaesthesia
North Bristol NHS Trust

Dr Andy Georgiou holds first-class honours in Anatomical Science and has demonstrated anatomy at the University of Bristol. He is now a consultant in anaesthesia and intensive care medicine at the Royal United Hospital in Bath.

Dr Chris Thompson holds the European Diploma in Regional Anaesthesia. He is a consultant in anaesthesia and a specialist in regional anaesthesia at North Bristol NHS Trust.

Dr James Nickells is a consultant in anaesthesia at North Bristol NHS Trust. Since 2004 James has been pursuing a second career as a professional artist and medical illustrator (www.jamesnickells.co.uk).

CAMBRIDGE
UNIVERSITY PRESS

CAMBRIDGE
UNIVERSITY PRESS

University Printing House, Cambridge CB2 8BS, United Kingdom

Cambridge University Press is part of the University of Cambridge.

It furthers the University's mission by disseminating knowledge in the pursuit of education, learning and research at the highest international levels of excellence.

www.cambridge.org
Information on this title: www.cambridge.org/9781107401372

© Andy Georgiou, Chris Thompson and James Nickells

First published 2014

Printed in Spain by Grafos SA, Arte sobre papel

A catalogue record for this publication is available from the British Library

Library of Congress Cataloguing in Publication data
Georgiou, Andy, author.
Applied anatomy for anaesthesia and intensive care / Dr Andy Georgiou,
Dr Chris Thompson, Dr James Nickells.
    p. ; cm.
Includes bibliographical references and index.
ISBN 978-1-107-40137-2 (pbk. : alk. paper)
I. Andy Georgiou 1978–, author.   II. Chris Thompson, author.
III. James Nickells, author.
[DNLM: 1. Anatomy. QS 4]
QM23.2
611–dc23

                                    2014001824

ISBN 978-1-107-40137-2 Paperback

....

Contents

Preface

Anatomical examination of the human body is a science which has been studied for over 3500 years. Anaesthesia and intensive care medicine are comparatively new specialties, evolving mostly in the twentieth century. However, it is only with the ready availability of high-quality portable ultrasound machines that the relevance and applicability of anatomy to these specialties has been brought so clearly into the working environments of the operating room and intensive care unit. The anatomical science once studied as an undergraduate now seems fundamental to our core knowledge and essential to improve the safety and quality of care delivered in both perioperative and critical care settings.

This book aims to distil years of anatomical study into a useful resource for the practising anaesthetist or intensive care specialist. By presenting only the most relevant anatomy in a concise and easy-to-read format, and by correlating it to the images gleaned from ultrasound, radiographic and fibreoptic technologies, we have produced an essential text for reference at the bedside, in the office and at home. However, in acknowledgement of the fact that not all readers will have access to, or be familiar with, such equipment, traditional procedural techniques such as those involving landmarks or nerve stimulator approaches are also described in full. Useful mnemonics and easily reproducible sketches of important anatomical areas are also provided for those studying towards postgraduate examinations in anaesthesia or intensive care medicine.

In summary, this text is an invaluable reference and study guide for practising anaesthetists and critical care physicians who wish to revise, develop or advance their anatomical knowledge and procedural skills, thereby increasing the scope, quality and safety of their practice.

Acknowledgements

From Andy:
To my wonderful family. To Lindsay for being so hugely patient and tolerant, and to Emily and Alex for bringing us so much joy.

From Chris:
To my wonderful family Hannah, Liberty and Charlie. For your patience and all the late nights.

From James:
To Theresa, Jasna, Kasia and Roxy. Thank you for your patience and understanding. To my mother. Thank you for inspiring creativity. To the memory of my dear father. Thank you for the work ethic.

Also to our friend and colleague Guy Matthew Jordan, 1972–2013.

Thanks to Drs Mike Coupe, Jerry Nolan, Tim Cook and Matthew Laugharne from the Royal United Hospital Bath. Thanks also to Drs Neil Rasburn, Matthew Molyneux and Kaj Kamalanathan from University of Bristol Hospitals NHS Foundation Trust.

We would also like to thank Deb Russell from Cambridge University Press for getting the project off the ground and Rob Walster at Big Blu for opening our eyes to the illustrative processes used in this text.

Disclaimer

Medicine is an ever-changing science. As new research and clinical experience broaden our knowledge, changes in treatment and drug therapy are required. The authors and the publisher of this work have checked with sources believed to be reliable in their efforts to provide information that is complete and generally in accord with the standards accepted at the time of publication. However, in view of the possibility of human error or changes in medical sciences, neither the authors nor the publisher, nor any other party who has been involved in the preparation or publication of this work, warrant all the information contained herein is in every respect accurate or complete, and they disclaim all responsibility for any errors or omissions or for the results obtained from use of the information contained in this work. Readers are encouraged to confirm the information contained herein with other sources. For example and in particular, readers are advised to check the product information sheet included in the package of each drug they plan to administer, to be certain that the information contained in this work is accurate and that changes have not been made in the recommended dose or in the contraindications for administration. This recommendation is of particular importance in connection with new or infrequently used drugs. Natural anatomical variation exists between individuals, and the authors cannot be accountable for the implication that this brings to clinical practice when using this text. The authors are not accountable for the actions of any person following the procedural guidance contained within this text. All persons must be appropriately trained by conventional means and be completely familiar with the procedures and equipment employed before performing any procedure or using any piece of equipment on patients, particularly when doing so in an unsupervised capacity. The risks and complications listed cannot be fully comprehensive, and the authors are not accountable for any untoward event that may arise and that is or is not listed in the text.

How to use this book

This book is designed to be used in two ways.

It is designed as a reference and core knowledge text for those working towards postgraduate examinations in anaesthesia and intensive care medicine. The book is organised by body region and contains useful mnemonics and easily reproducible sketch diagrams for key areas of the body. We hope it will also be a useful reference aid for those wishing to refresh this knowledge in the years after examinations.

It is also designed for those wishing to learn, revise and develop their procedural skills. All of the procedures are described in text boxes, which are located close to the anatomy relevant to that procedure.

Anatomical illustrations are correlated with ultrasound images where possible. The plane of the ultrasound beam is shown on the anatomical illustration by a grey 'pane' (akin to a pane of glass), showing how the structures are bisected by the ultrasound beam.

Please note, nerve blocks of the brachial plexus which are placed around or proximal to the clavicle are described in the neck chapter. Blocks placed distal to the clavicle are described in the upper limb chapter.

We hope you find the text useful and informative.

Universal procedural advice for peripheral nerve blockade

Multiple peripheral nerve block techniques are described in this book. Each technique contains some steps unique to that block and some which are universal. For brevity, and so that the reader can more rapidly identify the salient features of the specific block, universal pre- and post-procedure steps are listed here.

- Consent is required for all procedures.
- Procedures should be performed in a well-lit, clean environment.

Indications

Perioperative anaesthesia, postoperative analgesia and treatment of acute or chronic pain.

Absolute contraindications

- Patient refusal.
- Cutaneous infection at block insertion site.
- Allergy to local anaesthetic agent.

Relative contraindications

- Coagulopathy – see Table 1.
- Uncooperative patient.
- Patients with long-standing diabetes, peripheral neuropathy, sepsis or severe peripheral vascular disease may not experience nerve stimulation with a current intensity of < 0.5 mA, increasing the difficulty of locating the end point for injection.

Pre-procedure checks

- Airway and ventilation equipment.
- IV access.
- Resuscitation drugs and equipment.
- Trained assistant.
- Concious sedation and analgesia may be required to improve patient experience. Small titrated doses of fentanyl (up to 50 mcg) and propofol (target-controlled infusion or small boluses up to 50 mg) are effective.
- Supplementary oxygen.
- Aseptic technique. Meticulous skin preparation with 2% chlorhexidine gluconate in 70% isopropyl alcohol, allowed to dry fully. Sterile gloves and drapes.
- The ultrasound probe should be covered with a sterile probe cover.
- Nerve stimulator function should be checked prior to use – sufficient battery power, visual inspection of cables for damage and positive lead attached to the patient with an ECG gel sticker.
- 1–3 ml of local anaesthetic is infiltrated subcutaneously at the site of block needle insertion.

Complications

- Failure.
- Local anaesthetic toxicity.

Post-procedure checks

- Follow-up is required to establish block success and recovery.
- Care of the anaesthetised area/limb – protection from excessive heat, cold or trauma.

Anticoagulants and peripheral/central neuraxial blocks

The variety of anticoagulants being taken by patients is increasing. Judgement must always be exercised on the merits and risks of performing blocks in patients on anticoagulants (and if so when) or whether the drugs should be stopped prior to performing the block. Clearly stopping the drug requires an assessment of the underlying condition of the patient and an appreciation of the anaesthetic and surgical risks and benefits.

At time of writing, the following tables represent the most up-to-date information and advice on anticoagulants commonly in use, and they should help inform the reader on the benefits and risks of block performance in the anticoagulated patient.

Table 1 Recommendations related to drugs used to modify coagulation. Recommended minimum times are based in most circumstances on time to peak drug effect + (elimination half-life × 2), after which time < ¼ of the peak drug level will be present. For those drugs whose actions are unrelated to plasma levels, this calculation is not relevant. Data used to populate this table are derived from ASRA and ESRA guidelines [1,2] and information provided by drug manufacturers. These recommendations relate primarily to neuraxial blocks and to patients with normal renal function except where indicated.

Drug	Time to peak effect	Elimination half-life	Acceptable time after drug for block performance	Administration of drug while spinal or epidural catheter in place[1]	Acceptable time after block performance or catheter removal for next drug dose
Heparins					
UFH sc prophylaxis	< 30 min	1–2 h	4 h or normal APTTR	Caution	1 h
UFH iv treatment	< 5 min	1–2 h	4 h or normal APTTR	Caution[2]	4 h
LMWH sc prophylaxis	3–4 h	3–7 h	12 h	Caution[3]	4h[3]
LMWH sc treatment	3–1 h	3–7 h	24 h	Not recommended	4 h[4]
Heparin alternatives					
Danaparoid prophylaxis	4–5 h	24 h	Avoid (consider anti-Xa levels)	Not recommended	6 h
Danaparoid treatment	4–5 h	24 h	Avoid (consider anti-Xa levels)	Not recommended	6 h
Bivalirudin	5 min	25 min	10 h or normal APTTR	Not recommended	6 h
Argatroban	< 30 min	30–35 min	4 h or normal APTTR	Not recommended	6 h
Fondaparinux prophylaxis[5]	1–2 h	17–20 h	36–42 h (consider anti-Xa levels)	Not recommended	6–12 h
Fondaparinux treatment[5]	1–2 h	17–20 h	Avoid (consider anti-Xa levels)	Not recommended	12 h
Antiplatelet drugs NSAIDs	1–12 h	1–12 h	No additional precautions	No additional precautions	No additional precautions
Aspirin	12–24 h	Not relevant; irreversible effect	No additional precautions	No additional precautions	No additional precautions
Clopidogrel	12–24 h f		7 days	Not recommended	6 h
Prasugrel	15–30 min	8–12 h	7 days	Not recommended	6 h
Ticagrelor	2 h	4–8 h[6]	5 days	Not recommended	6 h
Tirofiban	< 5 min	4–8 h[6]	8 h	Not recommended	6 h
Eptifibatide	< 5 min	24–18 h[5]	8 h	Not recommended	6 h
Abciximab	< 5 min	10 h	48 h	Not recommended	6 h
Dipyridamole	75 min		No additional precautions	No additional precautions	6 h

			INR ? 1.4		After catheter removal
Oral anticoagulants Warfarin	3–5 days	4–5 days	INR ? 1.4	Not recommended	
Rivaroxaban prophylaxis[5] (CrCl > 30 ml.min⁻¹)	3 h	7–9 h	18 h	Not recommended	6 h
Rivaroxaban treatment[5] (CrCl > 30 ml.min⁻¹)	3 h	7–11 h	48 h	Not recommended	6 h
Dabigatran prophylaxis or treatment[7]					
(CrCl > 80 ml.min⁻¹)	0.5–2.0 h	12–17 h	48 h	Not recommended	6 h
(CrCl 50–80 ml.min⁻¹)	0.5–2.0 h	15 h	72 h	Not recommended	6 h
(CrCl 30–50 ml.min⁻¹)	0.5–2.0 h	18 h	96 h	Not recommended	6 h
Apixaban prophylaxis	3–4 h	12 h	24–18 h	Not recommended	6 h
Thrombolytic drugs					
Alteplase, anistreplase, reteplase, streptokinase	< 5 min	4–24 min	10 days	Not recommended	10 days

UFH, unfractionated heparin; sc, subcutaneous; APTTR, activated partial thromboplastin time ratio; iv, intravenous; LMWH, low molecular weight heparin, NSAIDs, non-steroidal anti-inflammatory drugs; INR, international normalised ratio; CrCl, creatinine clearance.

Notes to accompany Table 1

[1] The dangers associated with the administration of any drug that affects coagulation while a spinal or epidural catheter is in place should be considered carefully. There are limited data on the safety of the use of the newer drugs in this Table, and they are therefore not recommended until further data become available. The administration of those drugs whose entry in this column is marked as 'caution' may be acceptable, but the decision must be based on an evaluation of the risks and benefits of administration. If these drugs are given, the times identified in the column to the left ('Acceptable time after drug for block performance') should be used as a guide to the minimum time that should be allowed between drug administration and catheter removal.

[2] It is common for intravenous unfractionated heparin to be given a short time after spinal blockade or insertion of an epidural catheter during vascular and cardiac surgery. Local clinical governance guidelines should be followed and a high index of suspicion should be maintained if any signs attributable to vertebral canal haematoma develop.

[3] Low molecular weight heparins are commonly given in prophylactic doses twice daily after surgery, but many clinicians recommend that only one dose be given in the first 24 h after neuraxial blockade has been performed.

[4] Consider increasing to 24 h if block performance is traumatic.

[5] Manufacturer recommends caution with use of neuraxial catheters.

[6] Time to normal platelet function rather than elimination half-life.

[7] Manufacturer recommends that neuraxial catheters are not used.

References

1. Horlocker TT, Wedel DJ, Rowlingson JC, et al. Regional anesthesia in the patient receiving antithrombotic or thrombolytic therapy. *Reg Anesth Pain Med* 2010; **35**: 64–101.
2. Gogarten W, Van Aken H, Buttner J, et al. Regional anaesthesia and thromboembolism prophylaxis/anticoagulation: revised recommendations of the German Society of Anaesthesiology and Intensive Care Medicine. *Anasthesiol Intensivmed* 2007; **48**: S109–24.

Table 2 Relative risk related to neuraxial and peripheral nerve blocks in patients with abnormalities of coagulation.

	Block category	Examples of blocks in category
Higher risk ↑	Epidural with catheter Single-shot epidural Spinal	
	Paravertebral blocks	Paravertebral block Lumbar plexus block Lumbar sympathectomy Deep cervical plexus block
	Deep blocks	Coeliac plexus block Stellate ganglion block Proximal sciatic block (Labat, Raj, sub-gluteal) Obturator block Infraclavicular brachial plexus block Vertical infraclavicular block Supraclavicular brachial plexus block
	Superficial perivascular blocks	Popliteal sciatic block Femoral nerve block Intercostal nerve blocks Interscalene brachial plexus block Axillary brachial plexus block
	Fascial blocks	Ilio-inquinal block Ilio-hypogastric block Transversus abdominis plane block Fascia lata block
	Superficial blocks	Forearm nerve blocks Saphenous nerve block at the knee Nerve blocks at the ankle Superficial cervical plexus block Wrist block Digital nerve block Bier's block
Normal risk	Local infiltration	

Notes to accompany Table 2

There have only been 26 published reports of significant haemorrhagic complications of peripheral nerve and plexus blocks.[1] Half of these occurred in patients being given anticoagulant drugs and half in patients with normal coagulation. Patient harm has derived from:
• Spinal haematoma after accidental entry into the spinal canal during attempted paravertebral blocks as defined in the Table.
• Exsanguination.
• Compression of other structures, e.g. airway obstruction, occlusion of major blood vessels or tissue ischaemia.

The one death in this series was that of a patient on clopidogrel who underwent a lumbar plexus block and subsequently exsanguinated. The majority of the 26 cases underwent deep blocks or superficial perivascular blocks. From these data, and from other data relating to neuraxial blocks, we have placed blocks in the order of relative risk shown in the Table.

Catheter techniques may carry a higher risk than single-shot blocks. The risk at the time of catheter removal is unlikely to be negligible. Ultrasound-guided regional anaesthesia, when employed by clinicians experienced in its use, may decrease the incidence of vascular puncture, and may therefore make procedures such as supraclavicular blocks safer in the presence of altered coagulation.

1. Horlocker TT, Wedel DJ, Rowlingson JC, *et al.* Regional anesthesia in the patient receiving antithrombotic or thrombolytic therapy. *Reg Anesth Pain Med* 2010; **35**: 64–101.

Reproduced with permission from Harrop-Griffiths W, Cook T, Gill H, *et al.* Regional anaesthesia and patients with abnormalities of coagulation. *Anaesthesia* 2013; **68**: 966–72. doi: 10.1111/anae.12359.

Organisation of the nervous system

1. Subdivisions of the nervous system

The nervous system is divided:

i) Structurally

The central nervous system (CNS) – composed of the brain and spinal cord.

The peripheral nervous system (PNS) – composed of nervous tissue relaying information between peripheral structures and the CNS.

ii) Functionally

The somatic nervous system (SNS) – provides sensory and motor innervation to all parts of the body except the viscera, smooth muscle and glands.

The autonomic nervous system (ANS) – provides sensory and motor innervation to the viscera smooth muscle and glands.

2. Nervous tissue

Nervous tissue consists of:

i) Neurons

These are nerve cells.

They are composed of a nerve cell body, dendrites (projections from the cell body) and an axon (carries impulses to and from the cell body). Axons may be surrounded in myelin sheaths which are enclosed by endoneurium, a connective tissue layer.

The perineurium encloses groups of axons to form a fascicle.

The epineurium encloses groups of fascicles to form a peripheral nerve.

ii) Neuroglia

Non-excitable tissue.

Support the neurons structurally and metabolically.

3. Organisation of nervous tissue

Within the CNS, collections of nerve cell bodies form the grey matter; the interconnecting neural tissue forms the white matter.

Within the PNS, collections of nerve cell bodies form a ganglion; the interconnecting axons (held together by the epineurium) form a peripheral nerve.

Peripheral nerves may be cranial or spinal. There are:

- 12 pairs of cranial nerves, which arise from the brain and exit the cranium through the foramina of the skull.
- 31 pairs of spinal nerves (8 cervical, 12 thoracic, 5 lumbar, 5 sacral, 1 coccygeal), which arise from the spinal cord and exit through the intervertebral foramina.

4. Peripheral nervous system: anatomy of a spinal nerve

A typical spinal nerve originates from the spinal cord as dorsal (posterior) and ventral (anterior) rootlets which unite to form a dorsal and ventral root respectively (Figure 1.1).

- The dorsal fibres carry sensory (afferent) impulses to the dorsal horn of the spinal cord. The cell bodies of these nerves are found in the dorsal root ganglion.
- The ventral fibres carry motor (efferent) impulses from the ventral horn of the spinal cord. The cell bodies of these fibres are found in the grey matter of the ventral horn.

The dorsal and ventral roots unite to form a mixed spinal nerve (motor and sensory) at the intervertebral foramen, at which point the dural covering of the nerve ceases. Any divisions of the nerve from this point distally will therefore carry both motor and sensory impulses.

The spinal nerves of C1–7 emerge above their respective vertebrae. C8 spinal nerve emerges below C7 vertebra.

Moving inferiorly, all subsequent spinal nerves emerge below their respective vertebrae (T1 spinal nerve emerges below T1 vertebra and so on).

The spinal nerve then divides into dorsal and ventral primary rami.

- The dorsal primary ramus supplies the dorsal structures – the synovial joints of the vertebral column, musculature of the back and overlying skin.
- The ventral primary ramus supplies the ventral structures – the anterolateral trunk and the limbs.

Grey rami communicantes join the dorsal and ventral primary rami from the sympathetic trunk (see section 6, *Sympathetic nervous system*, below). The primary rami therefore carry somatic and autonomic, motor and sensory impulses.

The ventral rami of the spinal nerves in the cervical, brachial, lumbar and sacrococcygeal regions unite to form somatic plexuses which supply the neck, arms, legs and legs/pelvis respectively.

Autonomic plexuses are described in section 8, below.

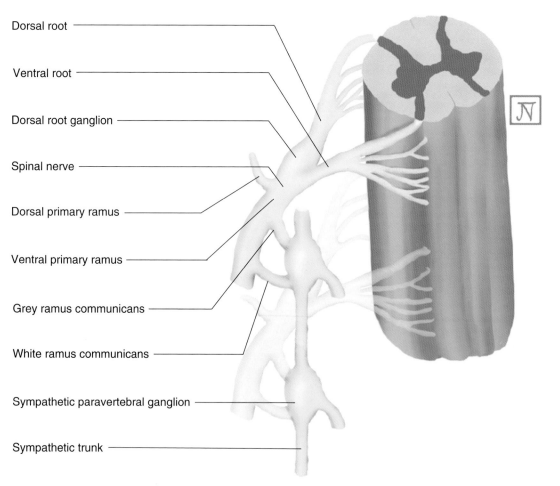

Dorsal root

Ventral root

Dorsal root ganglion

Spinal nerve

Dorsal primary ramus

Ventral primary ramus

Grey ramus communicans

White ramus communicans

Sympathetic paravertebral ganglion

Sympathetic trunk

Figure 1.1 The anatomy of a spinal nerve. The spinal nerve itself is located at the intervertebral foramen (the foramen formed between the pedicles of two adjacent vertebrae).

Exceptions to the above anatomy

- The first cervical dorsal primary ramus carries only motor impulses.
- The coccygeal dorsal primary ramus only supplies the skin over the coccyx.

The segmental arrangement of the primary rami leads to a segmental innervation of the overlying skin. The area of skin supplied by a single pair of spinal nerves is known as a dermatome (Figure 1.2). Note the absence of C1 dermatome.

Figure 1.2 The dermatomes. Note: there is no C1 dermatome.

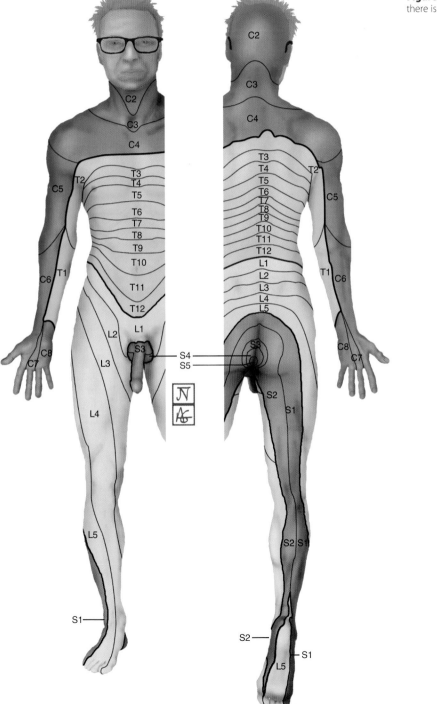

5. Organisation of the autonomic nervous system

Subdivided anatomically and functionally into the sympathetic and parasympathetic nervous systems. Both conduct impulses from the CNS via two neurons in series:

- The first (presynaptic or preganglionic) neuron has its cell body in the grey matter of the CNS and synapses only with the second neuron.
- The second (postsynaptic or postganglionic) neuron has its cell body in an autonomic ganglion located outside of the CNS. Its fibres terminate in the effector organs.

Neurotransmitters:

- Between the pre- and postganglionic neurons – acetylcholine.
- Between the postganglionic neuron and the effector – acetylcholine in the parasympathetic nervous system, noradrenaline in the sympathetic nervous system.

6. Sympathetic nervous system

The cell bodies of the sympathetic preganglionic neurons are found in the intermediolateral cell columns of the spinal cord. These columns are found in the lateral horns of the grey matter (Figure 2.7), between segments T1 and L2/3 of the spinal cord ('thoracolumbar' outflow).

The fibres leave the spinal cord through the ventral roots and enter the ventral primary rami.

Almost immediately, the fibres leave the ventral primary rami via white rami communicantes and enter the sympathetic trunk, which is a series of linked sympathetic *para*vertebral ganglia. There are 3 cervical (superior, middle and inferior cervical ganglia), 12 thoracic, 4 lumbar and 4 sacral ganglia within the sympathetic trunk (Figure 1.3). In 80% of people, the inferior cervical ganglion is fused with the first thoracic ganglion at the level of C7–T1 to form the stellate ganglion (described in more detail in Chapter 4, *The nerves of the neck*, section 3).

The preganglionic fibres then follow one of four pathways:

Pathway 1. Synapse with a postganglionic neuron at a paravertebral ganglion in the sympathetic trunk at the same spinal level.

Pathway 2. Ascend or descend within the sympathetic trunk before synapsing with a postganglionic neuron at a paravertebral ganglion at a higher or lower spinal level. This explains why, although the input to the sympathetic trunk comes from segments T1–L2/3, the trunk itself runs from C1 to the coccyx.

Pathway 3. Pass through the sympathetic trunk without synapsing. These neurons, known as the greater, lesser, least and lumbar splanchnic nerves, synapse with a postsynaptic neuron at one of the *pre*vertebral ganglia. The prevertebral ganglia, located about the origins of the branches of the abdominal aorta or in the pelvis, are:

- Coeliac ganglion – supplied by the greater splanchnic nerves (T5–9) and the lesser splanchnic nerves (T10–11).
- Superior mesenteric ganglion – supplied by the lesser splanchnic nerves (T10–11).
- Renal ganglion – supplied by the least splanchnic nerves (T12).
- Inferior mesenteric ganglion – supplied by the lumbar splanchnic nerves (L1–2).

Pathway 4. Fibres innervating the suprarenal gland also pass through the sympathetic trunk without synapsing. These fibres then pass through the coeliac prevertebral ganglion, again without synapsing, before innervating the medulla of the suprarenal gland. The suprarenal gland therefore acts as a special type of postganglionic neuron, releasing its neurotransmitter into the bloodstream instead of onto an effector organ.

Fibres supplying:

- The head, neck and limbs follow pathway 2 (ascending within the sympathetic trunk). More detail is supplied in Chapter 4, *The nerves of the neck*, section 3.
- The thorax follow pathway 1.
- The abdominal structures follow pathway 3.
- The pelvic structures follow pathway 2 (descending within the sympathetic trunk).

Postsynaptic fibres originating from the paravertebral ganglia (pathways 1 or 2) pass out of the sympathetic trunk through grey rami communicantes and re-join the ventral primary rami. From here, the fibres join all branches of the spinal nerve.

Postganglionic fibres originating from the prevertebral ganglia (pathway 3) form plexuses which track along the branches of the abdominal aorta to reach their effector sites (see section 8, *Autonomic plexuses*, below).

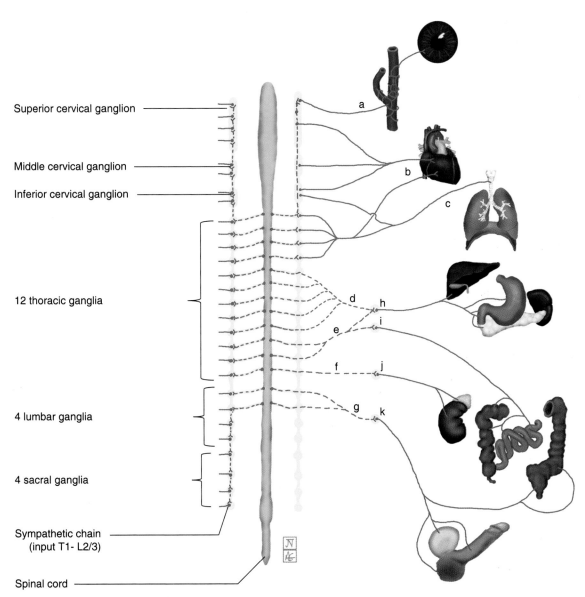

Superior cervical ganglion

Middle cervical ganglion

Inferior cervical ganglion

12 thoracic ganglia

4 lumbar ganglia

4 sacral ganglia

Sympathetic chain
(input T1- L2/3)

Spinal cord

Figure 1.3 Schematic anatomy of the sympathetic nervous system.

    a – Ascending sympathetic supply to head (follows arterial supply); follows pathway 2 (see text)
    b – Cardiac innervation, superficial cardiac plexus (left superior cervical ganglion) and deep cardiac plexus (all cervical ganglia (with the
        exception of the left superior cervical ganglion) and the upper four thoracic ganglia)
    c – Pulmonary innervation (via pulmonary plexus) inferior cervical ganglion and the upper four or five thoracic ganglia.
    d – Greater splanchnic nerve (T5–9)
    e – Lesser splanchnic nerve (T10–11)
    f – Least splanchnic nerve (T12)
    g – Lumbar splanchnic nerve (L1–2)
    h – Coeliac ganglion
    i – Superior mesenteric ganglion
    j – Renal ganglion
    k – Inferior mesenteric ganglion

The plexi (cardiac and pulmonary) also have parasympathetic innervation (not shown); the ganglia do not. The inferior cervical ganglion may be fused with the first thoracic ganglion to form the stellate ganglion (not shown for clarity).

7. Parasympathetic nervous system

The cell bodies of the presynaptic neurons are 'cranio-sacral' in origin.

i) Cranial outflow

These preganglionic fibres have their cell bodies in the grey matter of the brainstem. They leave the CNS in cranial nerves III, VII, IX and X and synapse with ganglia located in or on the walls of the effector organs (Figure 1.4). These ganglia are:

- Ciliary ganglion – receives CN III (pupillary constriction and accommodation).
- Pterygopalatine ganglion – receives CN VII (secretomotor to the lacrimal gland).
- Submandibular ganglion – receives CN VII (secretomotor to the submandibular and sublingual glands).
- Otic ganglion – receives CN IX (secretomotor to the parotid gland).

The vagus nerve (CN X) originates from three nuclei found in the medulla oblongata. It is responsible for the parasympathetic innervation to the thorax and abdominal viscera as far distally as the splenic flexure. Preganglionic fibres exit the skull through the jugular foramen. It forms several plexuses with the sympathetic nervous system (see section 8, *Autonomic plexuses*, below), before synapsing in ganglia in or on the walls of effector organs. The vagus nerve is described in more detail in Chapter 3, *The cranial nerves*, section 10.

ii) Sacral outflow

These preganglionic fibres have their cell bodies in the lateral horns of the grey matter of the spinal cord between S2 and S4. The fibres leave the CNS in the ventral roots of spinal nerves S2–4. The preganglionic fibres then leave the ventral rami as pelvic splanchnic nerves to synapse with ganglia in or on the walls of the descending and sigmoid colons and the inferior hypogastric plexus (see section 8, *Autonomic plexuses*, below). These fibres supply the pelvic viscera.

Given that the parasympathetic ganglia are on or close to the walls of the effector organs, it can be seen that the preganglionic fibres must be long and the postganglionic fibres short. This is in contrast to the sympathetic nervous system, where, given the para- or prevertebral location of the ganglia, the reverse is true.

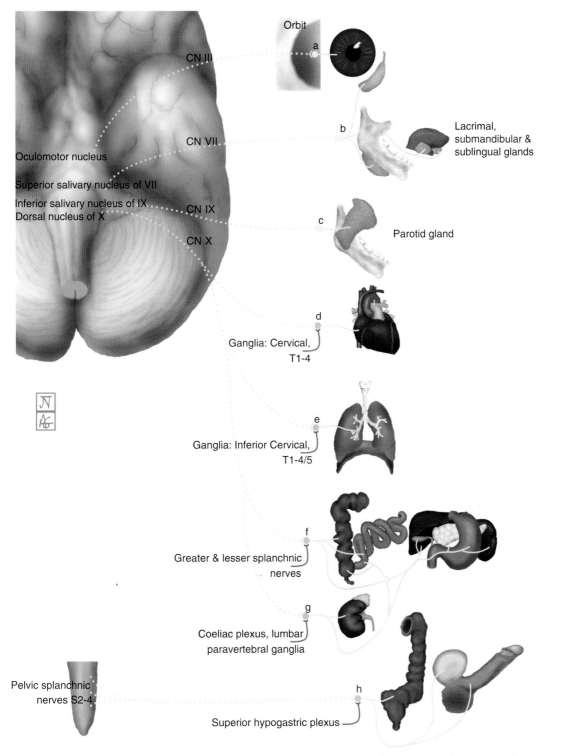

Figure 1.4 Schematic anatomy of the parasympathetic nervous system. Parasympathetic nerves are shown in light blue, sympathetic nerves in orange, ganglia in yellow and plexuses in green. The origin of the sympathetic fibres contributing to the autonomic plexuses is also shown.

a – Ciliary ganglion
b – Pterygopalatine ganglion (to the lacrimal glands) and submandibular ganglion (to the submandibular and sublingual glands)
c – Otic ganglion
d – Cardiac plexus
e – Pulmonary plexus
f – Coeliac plexus
g – Aortic plexus
h – Inferior hypogastric plexus

8. Autonomic plexuses

Autonomic plexuses receive sympathetic and para-sympathetic nerves from the pathways described above and distribute these to the viscera. The major plexuses are:

i) Superficial cardiac plexus

Found anterior to the pulmonary artery. It gives branches to the deep cardiac plexus and the pulmonary plexus.
Receives sympathetic innervation from the left superior cervical ganglion and parasympathetic innervation from the left vagus nerve.

ii) Deep cardiac plexus

Found anterior to the tracheal bifurcation. It innervates the heart.
Receives sympathetic innervation from the cervical ganglia (with the exception of the left superior cervical ganglion) and the upper four thoracic ganglia.
Receives parasympathetic innervation from the right vagus nerve.

iii) Pulmonary plexus

Found mostly posterior but also anterior to the roots of the lungs. Innervates the lungs and visceral pleura.
Receives sympathetic innervation from inferior cervical ganglion and the upper four or five thoracic ganglia.

Receives parasympathetic innervation from the vagus nerve, which synapses with postganglionic fibres in the plexus itself and those found along the walls of the bronchial tree.

iv) Coeliac plexus

Consists of the coeliac ganglia (see section 6, *Sympathetic nervous system*, above) and a series of interconnecting fibres of sympathetic and parasympathetic origin, which innervate the foregut.
There are two coeliac plexuses found to the left and right of the coeliac arterial trunk at its origin from the aorta at the level of L1. The relations of the plexuses are:

- Superior – crura of the diaphragm.
- Inferior – coeliac arterial trunk and its branches (hepatic, splenic and left gastric arteries).
- Anterior – inferior vena cava (on the right) pancreas and splenic artery (on the left).
- Posterior – abdominal aorta.

Receives preganglionic sympathetic fibres from the greater and lesser splanchnic nerves, which originate from thoracic segments T5–9 and T10–11 respectively. These nerves synapse with postganglionic fibres in the coeliac ganglion. It also receives the coeliac branch from the posterior vagal trunk, derived mainly from the right vagus nerve.

Sympathetic structures

Parasympathetic and mixed autonomic structures

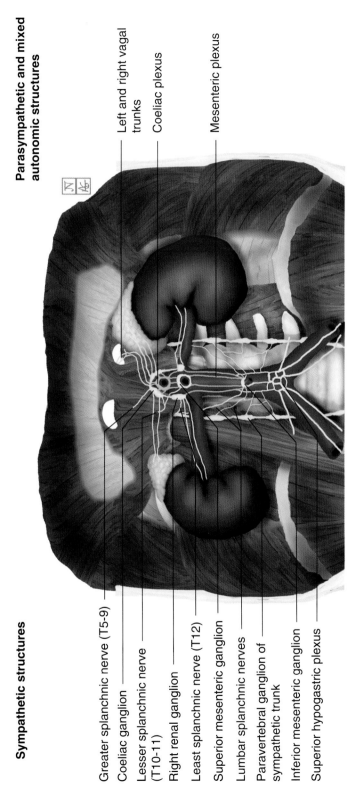

Greater splanchnic nerve (T5-9)

Coeliac ganglion

Lesser splanchnic nerve (T10-11)

Right renal ganglion

Least splanchnic nerve (T12)

Superior mesenteric ganglion

Lumbar splanchnic nerves

Paravertebral ganglion of sympathetic trunk

Inferior mesenteric ganglion

Superior hypogastric plexus

Left and right vagal trunks

Coeliac plexus

Mesenteric plexus

Figure 1.5 Autonomic anatomy of the abdomen. Sympathetic fibres are shown in yellow, parasympathetic fibres in blue, and where both nerves are found, green is used. The sympathetic trunk is a series of linked *para*vertebral ganglia where presynaptic sympathetic nerves synapse with postsynaptic sympathetic nerves (see text). The greater, lesser, least and lumbar splanchnic nerves (only the right-sided nerves are shown for clarity) pass through the sympathetic trunk without synapsing; they synapse with postsynaptic nerves at one of the *pre*vertebral ganglia – the coeliac, superior mesenteric, renal or inferior mesenteric ganglia. The suprarenal glands are supplied by nerves which pass through both the sympathetic trunk and the coeliac ganglion without synapsing. Parasympathetic innervation comes from the vagus nerve and the pelvic splanchnic nerves (not shown). Both sympathetic and parasympathetic fibres synapse at plexuses (shown in green): e.g. the coeliac plexus, which should be differentiated from ganglia as described above.

Coeliac plexus block

Introduction

Provides relief of pain carried by sympathetic nerves from the pancreas, stomach, liver, gall bladder, spleen, kidneys, small bowel and proximal two-thirds of the large bowel. It may be performed acutely during surgery for post-operative pain relief and electively for treatment of chronic pain and cancer pain, when an initial diagnostic local anaesthetic block may be followed by a neurolytic block.

Indications

Any condition causing severe upper abdominal visceral pain from the above organs that has failed to respond to standard treatments.

Specific contraindications

- Large aortic aneurysm.
- Where there is a possibility of seeding metastasis around the plexus.

Pre-procedure checks

Intravenous fluids are administered to reduce the degree of hypotension following the block due to dilation of the upper abdominal vessels.

Technique

X-ray screening.

Landmark

The patient is in the prone position. Local anaesthetic is infiltrated just below the tip of the 12th rib and the needle advanced, using x-ray screening in two planes, until it contacts the body of L1. The needle is then withdrawn slightly and redirected past the side of the vertebra forwards into the area of the coeliac plexus, avoiding the aorta and inferior vena cava. Radio-opaque dye is injected to confirm correct placement. A second injection from the other side is often required to achieve good spread.

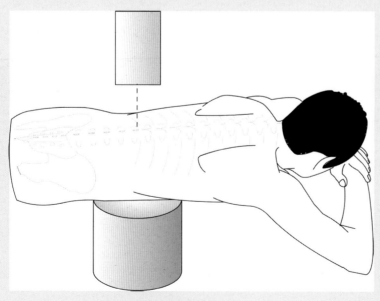

Figure 1.6 X-ray screening for a coeliac plexus block.

Ultrasound-assisted block

Endoscopic ultrasound-guided block is an emerging technique in which the ganglia are accessed anteriorly through the posterior stomach wall, thus avoiding major arteries, the diaphragm and the pleura. The coeliac plexus is not always seen as a discrete structure, but its location is determined relative to the coeliac trunk. A needle is introduced through the biopsy channel under real-time imaging, and aspiration is carried out to exclude vessel penetration prior to the injection being performed.

Complications

- Hypotension – may be severe even after unilateral block.
- Arteriovenous puncture and haemorrhage.
- Intravascular injection.
- Damage to upper abdominal organs causing cyst or abscess formation.
- Damage to the arterial supply to the spinal cord causing paraplegia.
- Nerve root irritation.

Post-procedure checks

- Chest x-ray to exclude pneumothorax.

v) Aortic plexus

The coeliac plexus descends over the abdominal aorta, forming the aortic plexus. Additional contributions to this plexus come from the lumbar paravertebral ganglia (see section 6, *Sympathetic nervous system*, above).

The plexuses then distribute along the branches of the aorta, taking their corresponding names (e.g. renal, testicular or ovarian plexuses) and supplying mixed autonomic innervation to the viscera.

vi) Hypogastric plexuses

The superior hypogastric plexus lies on the sacral promontory between the common iliac arteries. Receives sympathetic innervation from the aortic plexus and the lumbar splanchnic nerves (L1/2) and so is a sympathetic plexus without parasympathetic contribution.

These fibres then descend along the internal iliac arteries, where they are joined by the pelvic splanchnic nerves (supplying the parasympathetic innervation) to form the inferior hypogastric plexus, a mixed autonomic plexus (a paired structure).

The inferior hypogastric plexus delivers autonomic innervation to the pelvis.

9. Autonomic afferent fibres

Afferent impulses (such as visceral pain) may be conducted with sympathetic and parasympathetic nerves. However, these impulses do not synapse in autonomic ganglia. The synapses are in fact identical to those of the somatic nervous system (described in Chapter 2); the autonomic nerves act purely to convey the afferent information. For more information see Chapter 7, section 8 (*Visceral pain*).

The spine

The vertebral column

The vertebral column has four curvatures:

- The cervical and lumbar lordoses (concave posteriorly).
- The thoracic and sacral kyphoses (convex posteriorly).

Its function is to:

- Transmit weight through the pelvis.
- Protect the spinal cord.
- Provide a pivot for the head.
- Facilitate movement.

There are 33 vertebrae:

- 24 are true vertebrae (7 cervical, 12 thoracic, 5 lumbar).
- 9 are false vertebrae (5 sacral, 4 coccygeal).

1. True vertebrae

The features of a true (thoracic) vertebra are illustrated in Figure 2.2.

The pedicles and laminae together form the vertebral arch.

Consecutive vertebral foramina together form the vertebral canal, in which is found the spinal cord, meninges, epidural space and nerve roots.

The articulation of the superior and inferior facets results in the formation of the intervertebral foramen, out of which emerges the spinal nerve and dorsal root ganglion.

Anterior tubercle of C1

Dens of C2

Foramen of the transverse process

Superior articular facet (for articulation with the occipital condyles of the skull)

Transverse ligament of the atlas

Posterior arch and posterior tubercle

Dorsal primary ramus of C1

VF

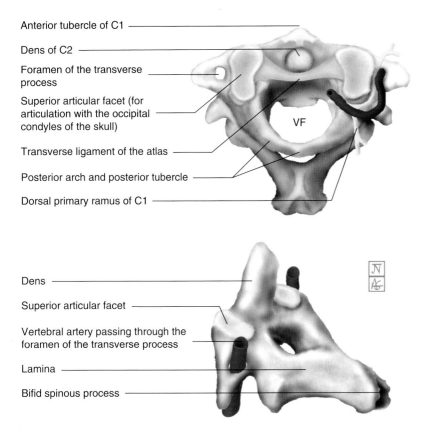

Dens

Superior articular facet

Vertebral artery passing through the foramen of the transverse process

Lamina

Bifid spinous process

Figure 2.1 The atlas (C1) and axis (C2). Upper figure: the articulation of the atlas and axis is shown. Note that the vertebral artery trends medially where it will unite with the artery from the left side to form the basilar artery. The spinal nerve of C1 is shown, with its division into the dorsal and ventral primary rami. VF, vertebral foramen. Lower figure: C2 in detail.

2. Cervical vertebrae

All but C7 have a foramen of the transverse process, which contains the vertebral arteries, venous and sympathetic plexuses.

i) C1 – the atlas

Consists of an anterior and posterior arch and a lateral mass.
It articulates with:
- The occipital condyles of the skull at its superior articular facets.
This articulation facilitates nodding of the head.
- The superior articular facets of C2 at its inferior articular facets.
- The dens of C2 at its facet in the anterior arch.

The latter two articulations facilitate rotation of the head.

ii) C2 – the axis

The dens (or odontoid peg) projects superiorly from its body. Articulation with C1 facilitates rotation of the head as described above.
The cruciate and alar ligaments maintain the position of the dens within the foramen of C1. The horizontal component of the cruciate ligament known as the transverse ligament of the atlas, importantly prevents posterior movement of the dens.

Vertebral Structures

Vertebral Joints

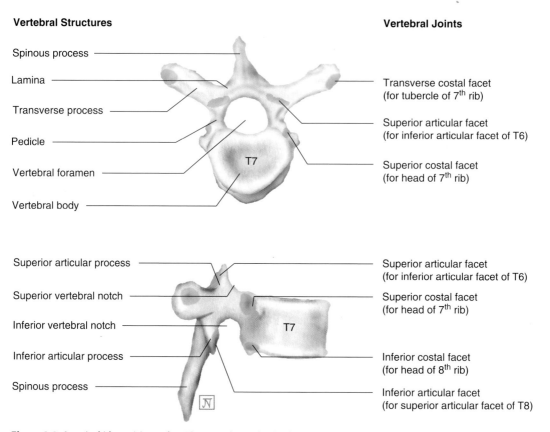

Spinous process

Lamina

Transverse process

Pedicle

Vertebral foramen

Vertebral body

T7

Transverse costal facet
(for tubercle of 7th rib)

Superior articular facet
(for inferior articular facet of T6)

Superior costal facet
(for head of 7th rib)

Superior articular process

Superior vertebral notch

Inferior vertebral notch

Inferior articular process

Spinous process

T7

Superior articular facet
(for inferior articular facet of T6)

Superior costal facet
(for head of 7th rib)

Inferior costal facet
(for head of 8th rib)

Inferior articular facet
(for superior articular facet of T8)

Figure 2.2 A typical (thoracic) vertebra. The seventh vertebra has been used to illustrate vertebral structures, and articulations with adjoining vertebrae and ribs. When adjacent vertebrae articulate, the superior and inferior vertebral notches align to form the intervertebral foramina, through which emerge the spinal nerves and dorsal root ganglia.

iii) C3–C7

The vertebral foramina are large (and roughly triangular) to accommodate the spinal cord in this area.

The transverse processes consist of anterior and posterior tubercles. The anterior tubercle of C6 is known as Chassaignac's tubercle.

The spinous processes are short and bifid except for C7, whose spinous process is large, making it the first cervical vertebra to be easily palpable (hence its name the 'vertebra prominens').

The foramen of the transverse process of C7 transmits only small veins.

Spinal nerves C1–7 emerge above their respective vertebrae.

Spinal nerve C8 emerges below C7 vertebra.

All subsequent spinal nerves emerge below their respective vertebrae.

3. Thoracic vertebrae

The vertebral bodies have costal facets for articulation with the head of the corresponding rib and the rib below (i.e. T6 vertebra articulates with the sixth and seventh ribs).

The transverse processes of vertebrae T1–10 have transverse costal facets for articulation with the neck of the corresponding rib (Figure 5.1).

The bodies are heart-shaped.

The spinous processes are long and the middle ones are directed inferiorly; the tip of the spinous process therefore aligns horizontally with the vertebral body below.

4. Lumbar vertebrae

All have:

- Large kidney-shaped bodies.
- Sturdy laminae.
- Absence of costal facets.
- Small transverse processes.

The body of L5 is wedge-shaped, being larger posteriorly than anteriorly, thus forming the lumbosacral angle (130–160°).

5. The sacrum

Formed by the fusion of five sacral vertebrae.
Articulates with:
- L5 vertebra superiorly.
- The ilium laterally (at the sacroiliac joints).
- The coccyx inferiorly.

Defining features:
- The sacral canal – the continuation of the vertebral canal in the sacrum. Contains epidural fat and veins and the meninges, within which are the cauda equina and filum terminale.
- Four pairs of sacral foramina – for exit of the dorsal and ventral spinal nerves S1–4.

Palpable posteriorly are:
- The median crest (the fused spinous processes of S1–4).
- The sacral hiatus (resulting from the absence of the laminae and spinous process of S5). The sacral hiatus is covered by the sacrococcygeal membrane (1–3 mm thick), closing the sacral canal.
- The sacral cornua (the inferior articular processes of S5).

Entry to the sacral canal (and therefore the caudal epidural space) is through the sacrococcygeal membrane at the sacral hiatus, which may be located in two ways:
- At the apex of an equilateral triangle whose base is formed by a line joining the posterior superior iliac spines (Figure 2.3).
- A triangular area bounded by the sacral cornua and the lowermost aspect of the median crest (formed by S4).

Caudal epidural block

Introduction

Caudal epidural anaesthesia is used predominantly in children, but also in adults for the management of acute and chronic pain. With advancing age, an increase in the fibrous content of the sacral epidural space (together with variation in the depth and width of the sacral canal) produces an inconsistent clinical effect.[1] The mean volume of the sacral canal is 33 ml (range 12–65 ml),[2] or 14.4 ml (range 9.5–26.6 ml) excluding the foramina and dural sac.[1]

Indications

Surgery of the perineum, penis and scrotum, rectum, bladder, inguinal and femoral regions.
Chronic pain use: lumbar radiculopathy or stenosis, failed back surgery syndrome and bony metastases.

Specific contraindications

- Unpredictable block height limits the usefulness of this technique in adults.

Pre-procedure checks

As for epidural.
In adults it is recommended that the block is performed with the patient awake, as the risk of local anaesthetic-induced seizures is greater than that for lumbar or thoracic epidurals.[3] The needle tip should stay below S2 to avoid the dura, the surface landmark of which is 1 cm below the level of the posterior superior iliac spines (PSIS).
In children the block may be performed with the patient anaesthetised. The needle tip should stay below S3/S4, as the dural sac extends further down the sacral canal in children.

Technique

Needle: 17G or 18G block needle or Tuohy.
An x-ray image intensifier may be useful when the landmarks are difficult to identify.

i) Landmark

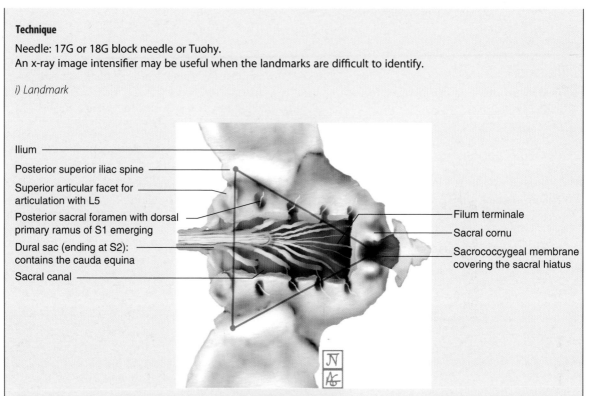

Ilium

Posterior superior iliac spine

Superior articular facet for articulation with L5

Posterior sacral foramen with dorsal primary ramus of S1 emerging

Dural sac (ending at S2): contains the cauda equina

Sacral canal

Filum terminale

Sacral cornu

Sacrococcygeal membrane covering the sacral hiatus

Figure 2.3 Sacral anatomy, viewed as if performing a caudal anaesthetic. A segment of the sacrum has been removed to reveal the relevant anatomy. The sacral canal is the continuation of the vertebral canal in the sacrum; it contains epidural fat and veins (not shown for clarity). The dural sac has been opened to reveal the cauda equina. The equilateral triangle used for identification of the sacrococcygeal membrane is shown (the base is formed by a line joining both posterior superior iliac spines).

The patient is in the lateral or prone position. An equilateral triangle is marked on the skin with the base formed by a line joining both PSIS and the apex pointing inferiorly to the sacral hiatus. The index finger is placed on the coccyx while the thumb palpates 3–4 cm cranially to find the two sacral cornua and the hiatus between them.

The needle is introduced through the sacral hiatus into the sacral canal. A 'pop' may be appreciated as the sacrococcygeal ligament is pierced. Alternatively, a loss-of-resistance technique may be employed. The needle is advanced until the ventral wall of the sacral canal is reached, then it is withdrawn slightly and redirected more cranially for insertion further into the canal.

Needle misplacement is suggested by bulging over the sacrum during injection (needle lies dorsal to the sacrum) or discomfort and resistance during injection (inadvertent puncture of the cortical layer of the sacral bone, mistaking this for the sacrococcygeal membrane).

ii) Ultrasound-assisted block

A low-frequency curved array probe is placed in a parasagittal plane 1 cm lateral to the spinous processes and angled towards the midline. The probe is moved caudally until the sacrum is identified (see box on *Epidural anaesthesia* – Figure 2.10a).

The probe is rotated into a transverse plane over the midline of the sacrum and moved caudally to identify the dorsal surface of the fifth sacral body and the sacrococcygeal membrane (Figure 2.4).

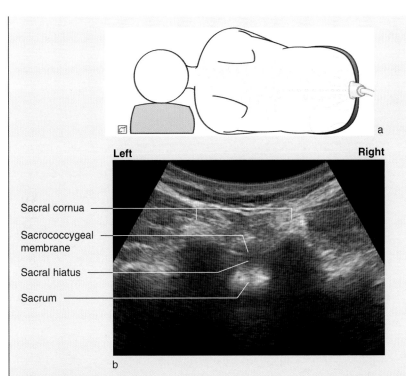

Left   Right

Sacral cornua —
Sacrococcygeal —
membrane
Sacral hiatus —
Sacrum —

b

Figures 2.4 Transverse ultrasound image of the approach to the caudal epidural space.

The sacral hiatus is fixed on the screen, the probe is rotated into a sagittal plane, and the needle is introduced in-plane (Figure 2.5).

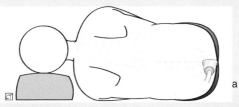

Superior   Inferior

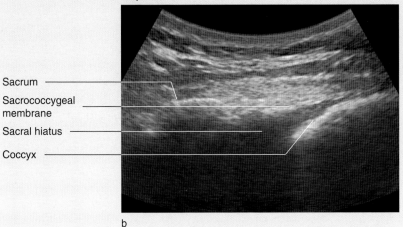

Sacrum —
Sacrococcygeal —
membrane
Sacral hiatus —
Coccyx —

b

Figures 2.5 Sagittal ultrasound image of the approach to the caudal epidural space.

Complications

As for epidural.
- Cauda equina syndrome – a syndrome of bowel and bladder dysfunction, patchy sensory deficit, pain and paresis of the legs.

Post-procedure checks

- The block should be assessed with light touch, pinprick or ice.

References

1. Crighton I, Barry B, Hobbs G. A study of the anatomy of the caudal space using magnetic resonance imaging. *Br J Anaesth* 1997; **78** 391–5.
2. Trotter M. Variations of the sacral canal: their significance in the administration of caudal anaesthesia. *Anesth Analg* 1947; **26**: 192–202.
3. Brown D, Ransom D, Hall J, *et al*. Regional anaesthesia and local anaesthetic-induced systemic toxicity: seizure frequency and accompanying cardiovascular changes. *Anesth Analg* 1995; **81**: 321–8.

6. The coccyx

Formed by the fusion of four small rudimentary coccygeal vertebrae.

Provides attachment for pelvic and gluteal muscles and ligaments.

The intervertebral discs and ligaments of the vertebral column

1. Intervertebral discs

Form secondary cartilaginous joints between the vertebral bodies and comprise up to 25% of the height of the vertebral column.

Composed of a tough outer annulus fibrosus and a soft, pulpy, avascular, elastic inner – the nucleus pulposus.

There is no intervertebral disc between C1 and C2 vertebrae.

2. Ligaments

From superficial to deep:

i) Suprasinous ligament

Tough cord-like ligament connecting the tips of the spinous processes from C7 to the sacrum.

Merges with the nuchal ligament, which extends from C7 to the posterior aspect of the foramen magnum.

ii) Interspinous ligaments

Connect the shafts of adjacent spinous processes.

iii) Ligamenta flava

(Singular – ligamentum flavum).

Thick elastic ligaments extending vertically between adjacent laminae.

Become thicker inferiorly; hence the elasticity is better appreciated with a Tuohy needle in the lumbar rather than the thoracic spine.

May become calcified in the elderly.

Prevent over-flexion of the spine.

iv) Posterior longitudinal ligament

Runs within the vertebral canal along the posterior aspect of the vertebral bodies and discs from C2 to the sacrum.

v) Anterior longitudinal ligament

Runs along the anterior surface of the vertebral bodies and discs from C2 to the sacrum.

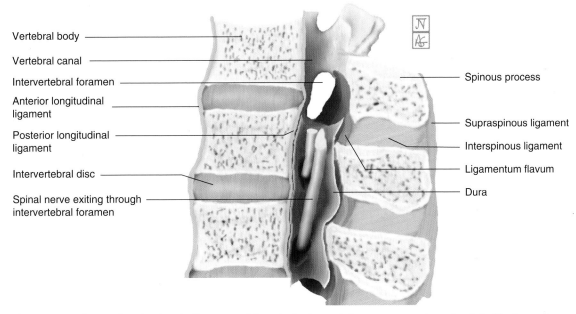

Vertebral body

Vertebral canal

Intervertebral foramen

Anterior longitudinal
ligament

Posterior longitudinal
ligament

Intervertebral disc

Spinal nerve exiting through
intervertebral foramen

Spinous process

Supraspinous ligament

Interspinous ligament

Ligamentum flavum

Dura

Figure 2.6 **Median section showing the ligaments of the spinal column.** Only two nerves are shown for clarity. The ligamenta flava are a series of tough elastic ligaments found between adjacent laminae; they are not one continuous ligament running the length of the spinal column.

The spinal cord and meninges

1. General structure

The spinal cord is a continuation of the medulla oblongata at the foramen magnum.

Differential growth of the cord and the vertebral column means that the cord ends at the lower border of L3 in the newborn and opposite the L1/2 intervertebral disc in the adult, attaining a final length of 42–45 cm. Individual variation means that the cord may terminate anywhere between the vertebral bodies of T12 down to L3 in the adult.

The differential growth means that the lumbar and sacral nerve roots must become elongated to reach their corresponding intervertebral foramina (where they exit the vertebral canal), thereby forming the cauda equina. So while the nerve roots of C8 leave the cord adjacent to the vertebral body of C7, the nerve roots of L5 leave the cord adjacent to the vertebral body of T12; they travel as the cauda equina before exiting at the L5/S1 intervertebral foramina.

The cord ends as the conus medullaris, which is attached to the coccyx by the filum terminale (Figure 2.3).

2. Gross cross-sectional anatomy

The cord is roughly circular in cross-section, being flattened in the anterior–posterior aspect.

The following structures are shown in Figure 2.7:

i) Central canal

Continuous with the fourth ventricle of the brain.

Contains cerebrospinal fluid (CSF), is lined with ciliated ependymal cells and is dilated in the region of the conus medullaris.

ii) Grey matter

Consists of neuronal cell bodies, glial cells and capillaries.

Forms an H-like structure, consisting of a transverse commissure and anterior and posterior columns (referred to as horns when viewed in transverse section), of which the anterior is the larger.

The posterior column has the substantia gelatinosa at its tip (Rexed lamina II).

The grey matter increases in size in the cervical and lumbar regions, corresponding to the area of limb innervation.

Descending Pathways

Ascending Pathways

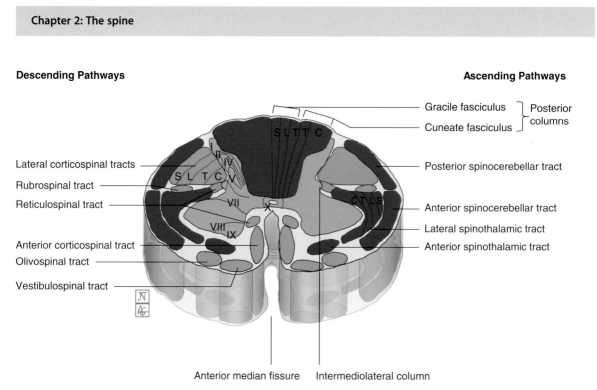

Figure 2.7 **Cross-section of the spinal cord at the mid-thoracic level showing ascending (purple) and descending (green) tracts.** Note the somatotopic organisation of the dorsal columns, lateral corticospinal tracts and anterior spinocerebellar tracts (C, cervical; T, thoracic; L, lumbar; S, sacral fibres). The central grey matter is organised into Rexed laminae, numbered sequentially from I to X, with X surrounding the central canal (not all numbers shown, for clarity).
Laminae I–VI = the dorsal horn; VII–IX = the ventral horn.
Lamina I = the marginal layer; II = the substantia gelatinosa, III–IV = the nucleus proprius.

A lateral grey column (or intermediolateral column) is found between T1 and L2/3, being the location of the cell bodies of the preganglionic neurons of the sympathetic nervous system.

iii) White matter

Consists of longitudinally disposed myelinated axon tracts (described in sections 3 and 4, below).
The amount of white matter decreases inferiorly as the number of afferent and efferent fibres joining or leaving the cord progressively decreases.

3. Descending fibre tracts

Relay efferent information.

i) Lateral corticospinal or pyramidal tract (pyramidal)

Motor tract carrying fibres to the limb muscles.
Upper motor neurons commence in the pyramidal cells of the motor cortex.
Fibres decussate in the medulla and descend in the tract.

Synapse with lower motor neurons found in the anterior horn (Rexed lamina VIII) at each spinal level.

ii) Anterior corticospinal tract (pyramidal)

Motor tract carrying fibres to axial muscles.
Upper motor neurons commence in the motor cortex.
Fibres descend in the tract.
At the appropriate spinal level, fibres decussate to synapse with the lower motor neurons of the contralateral anterior horn.

iii) Rubrospinal, tectospinal and vestibulospinal tracts (extrapyramidal)

Minor motor tracts.
Conduct extrapyramidal fibres passing from the brainstem nuclei to lower motor neurons, controlling:
- Rubrospinal – large muscle movement, inhibiting extension during flexing, for example.
- Tectospinal – reflex postural movements of the head, neck and eye in response to visual and auditory stimuli.

- Vestibulospinal tract – antigravity muscles, muscles controlling head and neck position and gaze.

4. Ascending fibre tracts

Relay afferent information.

i) Posterior (dorsal) columns

Carry information on fine touch, vibration and proprioception to the contralateral cerebrum. Divided into the gracile (afferents from T7 and below) and cuneate (afferents from T6 and above) fasciculi.

First-order neurons ascend to synapse with second-order neurons in the gracile and cuneate nuclei respectively, in the medulla.

Second-order neurons decussate and pass in the medial lemniscus to the ventral posterolateral nucleus of the thalamus, where they synapse with the third-order neurons.

The third-order neurons conduct impulses to the sensory cortex (although some fibres pass to the cerebellum).

ii) Spinothalamic tracts

Carry information on pain, temperature, crude touch and pressure to the contralateral cerebrum.

First-order neurons synapse with second-order neurons in the posterior horn of the spinal cord (in laminae I, II and V).

The second-order neurons decussate at that spinal level and ascend in the lateral (afferents for pain and temperature) or anterior (afferents for crude touch and pressure) spinothalamic tracts. They synapse with the third-order neurons at the ventral posterolateral nucleus of the thalamus where third-order neurons pass the postcentral gyrus of the cortex (some fibres pass to the reticular formation, allowing a painful stimulus to elevate alertness).

iii) Spinocerebellar tracts

Carry proprioceptive information to the ipsilateral cerebellum.

First-order neurons synapse with second-order neurons in the posterior horn of the spinal cord (in laminae I–VI).

Second-order neurons ascend in the posterior (from the upper body) and anterior (from the lower body) spinocerebellar tracts.

The former remain uncrossed and synapse with the third-order neurons at the inferior cerebellar peduncle. The latter decussate, first at the original spinal level and again prior to synapsing with the third-order neurons at the superior cerebellar peduncle (hence both pathways convey ipsilateral information).

Third-order neurons convey information to the cerebellum.

Ascending and descending pathways display somatotopic organisation – fibres to or from adjacent parts of the body are found adjacent to one another in the central nervous system (Figure 2.7), allowing a prediction of neurological deficit in pathological situations. Ascending and descending fibres in the lateral part of the cord white matter innervate distal limb muscles; those in the medial portion innervate the axial muscles and the upper limbs. The posterior columns also show somatotopic organisation as outlined above.

5. The anatomy of pain

Pain is 'an unpleasant sensory and emotional experience associated with actual or potential tissue damage or described in terms of such damage'.

i) First-order neurons

Free nerve endings (nociceptors) in skin, subcutaneous tissue, periosteum, joints, muscles and viscera are activated by chemical, thermal, or mechanical energy or by physical disruption of the tissue.

Action potentials are then conducted in myelinated Aδ (2–5 μm diameter, 10–30 m/s velocity) or unmyelinated C (0.5–1 μm diameter, 0.5–2 m/s velocity) fibres.

These first-order neurons have their cell bodies in the dorsal root ganglia and enter the dorsal horn in an area known as Lissauer's tract.

The fibres terminate by synapsing with second-order neurons: Aδ fibres do so in laminae I, II, V and X; C fibres do so in laminae I, II and V (Figure 2.7).

ii) Second-order neurons

Second-order neurons then decussate to transmit nociceptive action potentials: 90% travel via the lateral (pain and temperature) and anterior (crude touch and pressure) spinothalamic tracts to the thalamus; 10% travel in the spinoreticular tract to

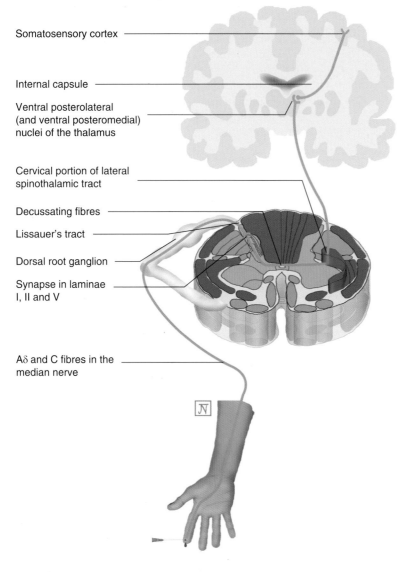

Somatosensory cortex

Internal capsule

Ventral posterolateral
(and ventral posteromedial)
nuclei of the thalamus

Cervical portion of lateral
spinothalamic tract

Decussating fibres

Lissauer's tract

Dorsal root ganglion

Synapse in laminae
I, II and V

Aδ and C fibres in the
median nerve

Figure 2.8 Schematic representation of the pain pathway from the upper limb. Only the major pathways essential for transmission of a painful stimulus are shown, for clarity.

the reticular formation in the brainstem. The cell bodies of the former pathway are found in laminae I and V, and of the latter pathway in laminae VII and VIII.

The spinothalamic tracts project to the hypothalamus, the periaqueductal grey and the reticular formation of the medulla before terminating in the ventral posterolateral (mostly) but also ventral posteromedial nuclei of the thalamus. They are concerned with the quantitative and spatial components of the pain signal.

The spinoreticular tract is concerned with the more primitive reflex and endocrine responses associated with a painful stimulus.

iii) Third-order neurons

Third-order neurons project from the thalamic nuclei to the somatosensory cortex via the internal capsule. This part of the brain is responsible for the discriminatory elements of painful stimuli, but the global appreciation of it comes from other parts of the brain such as the cingulate gyrus, the thalamus and the frontal lobes.

iv) Modulation of pain

The pain signal may be innately modulated:
- Centrally, particularly by the periaqueductal grey in the midbrain.
- Spinally by descending inhibitory pathways (noradrenergic and serotonergic).
- Spinally by simultaneous gated inputs (gate control theory).

Further description of the modulation of nociceptive inputs is beyond the scope of this text.

6. Spinal meninges

The meninges (from outermost to innermost) are the dura mater, arachnoid mater and pia mater. They surround, support and protect the spinal cord and are in continuation with the meninges surrounding the brain (Chapter 3, *The brain*, section 3).

i) Dura mater

The continuation of the inner layer of dura surrounding the brain.

It is a tough, fibrous layer which is separated from the vertebrae by the epidural space.

It is anchored to the margin of the foramen magnum superiorly, by the filum terminale (which it ensheaths) inferiorly, and by slender filaments to the posterior longitudinal ligaments anteriorly.

It is, however, unattached posteriorly.

It extends through the intervertebral foramina (forming dural root sleeves) and passes along the dorsal and ventral nerve roots distal to the spinal ganglia.

It ends by fusing with the epineurium of the spinal nerves.

The dural sac extends inferiorly as far as S2, but may end as high as L5 or as low as S3 (where it may be at risk of puncture during caudal anaesthesia).

ii) Arachnoid mater

A delicate avascular membrane which lines the dura and the dural root sleeves.

It is not attached to the dura but is held against it by the pressure of the CSF.

iii) Pia mater

Vascular connective tissue that closely invests the surface features of the spinal cord and covers the roots of the spinal nerves.

It attaches to the dura laterally via strands known as denticulate ligaments, and posteriorly by the incomplete posterior subarachnoid septum.

It continues as the filum terminale inferiorly.

7. Meningeal spaces

i) Epidural or extradural space

This refers to the contents of the spinal canal found external to the dura.

It extends from the foramen magnum to the sacrococcygeal membrane.

The space is roughly triangular in cross-section, with the base anterior (where it is very narrow) and the apex posterior, where it may be 6 mm deep in the lumbar region and 1 mm deep in the cervical region.

Boundaries:
- Internal – dura mater of spinal cord.
- Posterior – ligamenta flava and periosteal lining of the vertebral laminae.
- Anterior – posterior longitudinal ligament.
- Laterally – intervertebral foramina and periosteal lining of the vertebral pedicles. The space does extend for a short distance laterally through the intervertebral foramina where it accompanies the nerve roots.

Contents:
- Epidural fat.
- Epidural blood vessels. The veins (Batson's plexus) from a valveless communication between the pelvic and cerebral veins, explaining the engorgement which occurs during uterine contraction in pregnancy (increasing the chance of a 'bloody tap') and forming a pathway for the spread of malignant cells or bacteria. Shunting of blood to the epidural veins during coughing accounts for the raised CSF pressure seen.
- Lymphatics.
- Spinal nerve roots.
- Connective tissue, stretching from the dura to the ligamenta flava, which may divide the epidural space and explain poor (or even unilateral) distribution of anaesthetic agents in some patients.

The pressure in the epidural space is negative, because of the transmission of negative intrapleural pressure through the paravertebral spaces and intervertebral foramina. An additional artefactual component comes from tenting of the dura by the needle and possibly recoil of the ligamentum flavum following its puncture.

The usual distance between the skin and the epidural space is 2–9 cm.

ii) Subdural space

A potential space, closed by the pressure of CSF pressing the arachnoid against the dura.

The eye of a Whitacre needle may, in rare circumstances, straddle this space, and inadvertent deposition of local anaesthetic solution into it would result in inadequate spinal anaesthesia. Alternatively an epidural catheter placed inadvertently into it would result in extensive but patchy anaesthesia.

iii) Subarachnoid space

An actual space found between the arachnoid and the pia mater and containing CSF.

Approximately 20 ml of CSF is found in the spinal subarachnoid space; the remaining 100 ml is found within the skull.

It is enlarged in the dural sac inferior to the conus medullaris, where it forms the lumbar cistern, the target for spinal anaesthetic injections or lumbar punctures.

8. Vascular supply

i) Arterial

- One anterior spinal artery – formed by the union of branches from each vertebral artery at the foramen magnum. It runs down the anterior median fissure, supplying the anterior two-thirds of the cord.
- Two posterior spinal arteries – derived from the posterior inferior cerebellar arteries. They descend on the posterolateral aspect of the cord and supply the posterior one-third of it.
- Radicular arteries – arise locally (e.g. deep cervical, intercostal, lumbar) and supply additional flow to the three spinal arteries. The biggest, the artery of Adamkiewicz (arteria radicularis magna) arises in the lower thoracic/upper lumbar region (on the left in 65% of people) and may be responsible for the circulation to the lower two-thirds of the spinal cord.

There are no anastomoses between the anterior and posterior spinal arteries. The perfusion to the cord, particularly between T3–5 and T12–L1, is therefore vulnerable during low flow states such as hypotension, surgical occlusion and vasoconstriction.

ii) Venous

- Three anterior and three posterior spinal veins drain to the vertebral venous plexus in the epidural space.
- From here, veins drain to the dural venous sinuses – the vertebral, azygous or lateral sacral veins, depending on the spinal level.

Spinal cord

Dural cuff enveloping the dorsal and ventral roots (not seen) and the spinal nerve

2 posterior spinal arteries

Arachnoid mater

Dura mater

Epidural space filled with fat and vessels

Ligamentum flavum

Lamina

Spinous process

Transverse process

Rib

Interspinous ligament

Supraspinous ligament

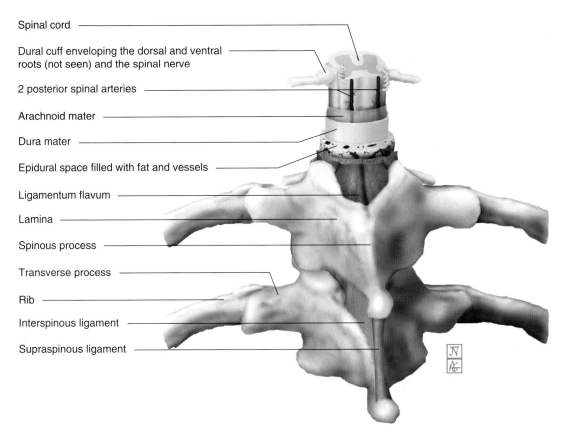

Figure 2.9 The anatomy of the approach to the thoracic epidural space. The epidural space is defined by the contents of the spinal canal external to the dura, and is seen here between the ligamentum flavum and the dura where it is filled with epidural fat and vessels (particularly veins).

Epidural anaesthesia

Introduction
Epidural blockade is a safe and effective means of providing analgesia during labour or following surgery. It may be used as a sole anaesthetic technique for some operations.

Indications
Anaesthesia and analgesia typically for procedures involving the thorax, abdomen, lower limbs, pelvis, perineum, and during labour. Patients in whom general anaesthesia may lead to prolonged ventilation, such as those with lung disease, may benefit from a thoracic epidural.

Specific contraindications
Absolute:
- Patient refusal.
- Cutaneous infection at block insertion site.
- Severe uncorrected hypovolaemia.
- Raised intracranial pressure.
- Undefined or unstable neurological disease.
- Allergy to local anaesthetic agent (rare).

Relative:
- Uncooperative patient.
- Abnormalities of the vertebral column.
- Coagulopathy – see *Universal procedural advice*, Table 1.
- Significant cardiac disease (e.g. severe aortic stenosis, mitral stenosis, hypertrophic cardiomyopathy) where a reduction in afterload or preload represents significant risk. Blocks above T5 have a greater effect on patient haemodynamics than those at T10 or below, due to the extent of sympathetic block and the loss of sympathetic innervation to the heart.
- Drugs that influence the patient's response to sympathetic blockade, e.g. alpha-blocking agents.
- Myasthenia gravis, pulmonary fibrosis and severe chronic obstructive disease require careful preoperative evaluation but do not necessarily exclude a patient.

Pre-procedure checks

- Airway and ventilation equipment.
- IV access.
- Vasopressors and inotropes.
- Resuscitation drugs and life support equipment.
- Monitoring including pulse oximetry and blood pressure – as recommended by the AAGBI minimum standards.
- Trained assistant.
- Sedation and analgesia may be required.
- Supplementary oxygen.
- Full aseptic technique (gown/mask/gloves/hat). 0.5% chlorhexidine gluconate in 70% isopropyl alcohol is recommended for skin preparation.[1] It should be allowed to dry fully, and care must be taken not to contaminate the epidural needle and catheter.

Technique

Needle: 16G Tuohy with 1 cm depth markings.

Landmark

An appropriate level is determined using surface anatomical landmarks:
- C7 spinous process – the most prominent process in the neck.
- T7 spinous process – level with the inferior angle of the scapula.
- L4 spinous process – bisected by Tuffier's line, drawn between the superior aspects of the iliac crests.

The sitting position allows easy identification of the midline and surface anatomy, particularly in obese patients. The patient sits on the bed with feet on a stool, arms over a pillow, shoulders relaxed, chin on the chest and with the back maximally flexed to open up the vertebral spaces. An assistant is useful to help the patient maintain his or her position. The lateral position is commonly used if sedation is required, in obstetrics, and if an assistant is not available to help the patient maintain position. The patient is placed on his or her side with the back at the edge of the bed, knees drawn up to the chest and chin on chest.

Midline approach

The midline is identified by palpating the spinous processes. Local anaesthetic is infiltrated into the skin at the midpoint between two adjacent vertebrae and then deeper into the tissues to confirm the location of the midline. The epidural needle is inserted through the same skin puncture, with the back of the non-injecting hand resting on the patient's back, bracing the hub of the epidural needle. The needle is inserted about 2–3 cm until it is gripped firmly by the supra- and interspinous ligaments. The stylet is then removed and a low-resistance syringe filled with saline is attached to the hub of the needle. Continuous pressure is exerted on the plunger of the syringe and the needle advanced until the resistance to injection of saline falls (loss-of-resistance technique). The depth is noted, the syringe is removed, and the non-injecting hand is used to hold the needle in place.

The epidural catheter is fed through the needle up to the 15–17 cm mark, then the needle is removed without dislodging the catheter. The catheter is withdrawn to the skin-to-epidural depth plus 5 cm – leaving 5 cm of catheter in the epidural space. The catheter is aspirated to detect intravascular or intrathecal placement and flushed with a small amount of saline to ensure patency. The fluid level within the catheter should fall. A clear, occlusive dressing is then applied over the insertion site.

Paramedian approach
This approach presents a larger opening into the epidural space. It is useful when the patient cannot easily flex the spine, for a calcified interspinous ligament, and for mid-thoracic epidurals where the angulation of the spinous processes is greater. Local anaesthetic is infiltrated 2 cm lateral to the midline at the level of the interspace in the lumbar region, or 2 cm lateral to the spinous process in the mid-thoracic region. The needle is inserted perpendicularly and slightly cephalad until the lamina is contacted. The needle is withdrawn slightly, then redirected 20° cephalad and towards the midline. If bone is encountered the needle is 'walked off' the bone into the ligamentum flavum and the loss-of-resistance technique is applied as above. The remaining steps are as described above.

Ultrasound-assisted block

Evidence on ultrasound-guided catheterisation of the epidural space is limited, but NICE Interventional Procedure Guidance 249 suggests that it is safe and may be helpful to achieve pre-procedure identification of the midline, level and depth of the space.[2] Real-time ultrasound placement of epidural catheters has been advocated, particularly in children, but the technique is cumbersome.

1) Identify the sacrum
A low-frequency curved array probe is placed in a parasagittal plane 1 cm lateral to the spinous processes and angled towards the midline. The sacrum is identified and the L5/S1 interspace level marked with a skin marker lateral to the midline.

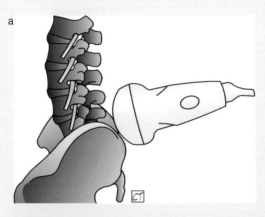

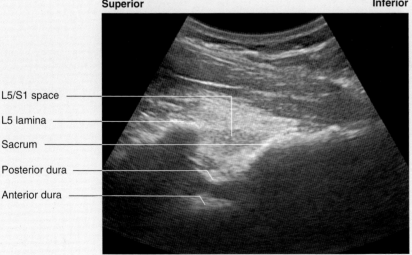

Figure 2.10 Sagittal ultrasound images illustrating the method for locating the correct vertebral interspace. The sacrum and L5/S1 interspace are identified initially.

2) Mark the vertebral level

Using L5/S1 as the starting point, additional interspace levels are identified sonographically and marked with a skin marker lateral to the midline.

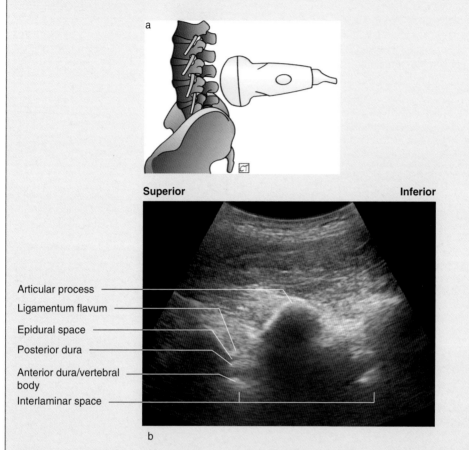

Figure 2.11 Sagittal ultrasound images illustrating the method for locating the correct vertebral interspace. The probe is moved superiorly through the vertebral levels until the desired space is reached.

3) Mark the midline and note the angle of insertion

The probe is placed in a transverse plane and the bright reflections of the spinous processes are identified. The midpoint of each process is marked.

The probe is tilted to look between the spinous processes. As the beam passes through the interspace, the bright reflection of the ligamentum flavum and ventral dura can be seen. The angle of tilt of the probe is noted, so that the needle can be inserted in the same plane. The plastic cover from an 18G needle may be useful to indent the skin at the insertion point, as the skin marker may be washed off by the chlorhexidine skin preparation.

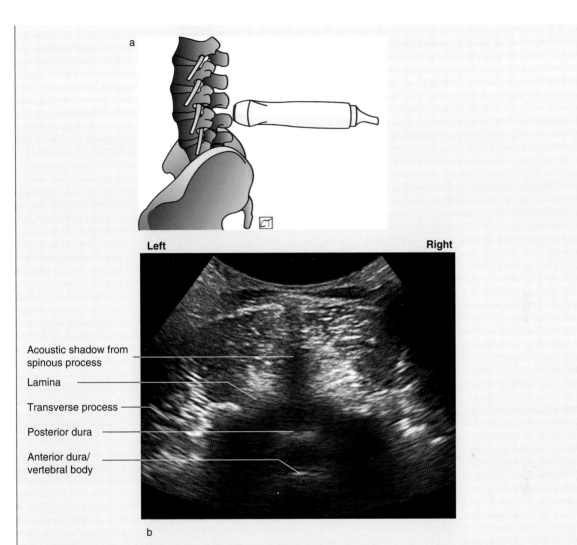

Acoustic shadow from
spinous process

Lamina

Transverse process

Posterior dura

Anterior dura/
vertebral body

Figure 2.12 Transverse ultrasound image of the intervertebral space. The midline is located, the angle of insertion is noted, and the depth to the epidural space is measured.

4) Measure the depth
The image is frozen and the on-screen callipers are used to measure the insertion depth. The ultrasound depth may differ from the needle insertion depth. This may be due to the pressure exerted by the probe compressing the tissues during scanning. The gel is completely removed from the patient's back and the epidural is inserted as described above.

Complications

A UK national audit of central neuraxial blocks detected 52 major complications in an estimated 700,000 procedures.[3] Two-thirds of patients with complications made a full recovery. Where permanent harm occurred, 60% did so after epidural block and 23% after spinal anaesthesia. Complications included:

- Minor back pain – 20–30% after epidural.
- Postdural puncture headache – 1%.
- Permanent nerve injury – 1 in 24,000 to 1 in 54,000.
- Paraplegia or death – 1 in 50,000 to 1 in 140,000.
- Spinal cord/nerve root injury – due to direct trauma, spinal cord ischaemia, neurotoxic drugs, haematoma or abscess.

- Accidental subdural or subarachnoid injection.
- Blood vessel puncture – more common in pregnant patients.
- Meningitis.
- Chronic adhesive arachnoiditis.
- Inadvertent intravenous injection of local anaesthetic.

Post-procedure checks

- A test dose is given to detect intravascular or subarachnoid placement. The dose should be less than 12.5 mg bupivacaine, i.e. a safe intrathecal dose.
- Volume is the key factor in determining the height of the block. 1–2 ml per segment to be blocked is a generally accepted guideline.
- A decrease in blood pressure is common and expected. Appropriate fluids, vasopressors or inotropes are administered.
- The block should be assessed with light touch, pinprick or ice 10–15 minutes later and additional boluses given as required.
- Unintentional injection into the subdural space may produce a diffuse, patchy epidural with a delayed onset, or a high spinal.

References

1. Cook T, Fischer B, Bogod D, *et al*. Correspondence. Antiseptic solutions for central neuraxial blockade: which concentration of chlorhexidine in alcohol should we use? *Br J Anaesth* 2009; **103**: 456–7.
2. National Institute for Health and Clinical Excellence. *Ultrasound-guided catheterization of the epidural space*. Interventional Procedure Guidance 249. London: NICE; 2007. http://www.nice.org.uk/guidance/IPG249 (accessed November 2013).
3. Cook T, Counsell D, Wildsmith J, *et al*. Major complications of central neuraxial block: report on the Third National Audit Project of the Royal College of Anaesthetists. *Br J Anaesth* 2009; **102**: 179–90.

Spinal anaesthesia

Introduction

Spinal anaesthesia is useful either as the sole anaesthetic technique or in combination with general anaesthesia for peri- and postoperative analgesia.

Indications

The level of dermatomal block required varies according to the surgical procedure performed:
- T4 – Caesarean section (traction on the peritoneum and exteriorisation of the uterus) or upper abdominal surgery.
- T6 – intestinal (not as sole anaesthetic), gynaecological and urological surgery.
- T10 – transurethral resection of the prostate, vaginal delivery or hip surgery.
- L1 – thigh or lower leg surgery.
- L2 – foot and ankle surgery.
- S2–5 – perineal and anal surgery (saddle block).

Specific contraindications

See box on *Epidural anaesthesia*.
- Use of lidocaine is associated with transient neurologic symptoms (TNS) including low back pain and lower limb dysaesthesia.[1]

Pre-procedure checks

See box on *Epidural anaesthesia*.

Technique

Needle: 25G non-cutting, pencil-point spinal needle (Sprotte or Whitacre) with a removable stylet.

Landmark

See box on *Epidural anaesthesia*.
The L4 spinous process is bisected by Tuffier's line, drawn between the superior aspects of the iliac crests. Spinal anaesthesia is commonly performed at the L3–4 interspinous space, as the spinal cord ends at the L1–2 vertebral level in adults and at the L3 vertebral level in children.

Midline approach

The midline is identified by palpating the spinous processes. Local anaesthetic is infiltrated into the skin at the midpoint between two adjacent vertebrae and then deeper into the tissues to confirm the location of the midline. The introducer needle is inserted through the same skin puncture with a 15° cephalad angle. The spinal needle is passed through the introducer and the subcutaneous tissue, followed by the ligaments noted in the text, prior to breaching the dura and arachnoid mater and entering the subarachnoid space. A 'pop' is often felt as the needle goes through the dura mater. The stylet is removed to check for CSF, and local anaesthetic is injected.

Paramedian approach

This approach is useful when the patient cannot easily flex the spine or has a calcified interspinous ligament. The needle is inserted 1 cm lateral and 1 cm inferior to the spinous process of the level required, and directed towards the middle of the interspace. If the lamina is contacted the needle is 'walked off' the bone into the ligamentum flavum. The ligamentum flavum is often the first resistance felt before reaching the subarachnoid space.

Ultrasound-assisted block

See box on *Epidural anaesthesia*.
The same technique described for epidural placement may be used for pre-procedure ultrasound-assisted marking of the midline, level and depth prior to spinal insertion.

Dose

The dose of local anaesthetic and adjunctive drugs used depends on the block height and duration required. A typical dose for Caesarean section is 2.5 ml 0.5% heavy bupivacaine with 300 mcg diamorphine, whereas a total knee replacement may need 3 ml of 0.25% bupivacaine.
The three most important factors affecting block height are: baricity of the local anaesthetic, patient position during and just after injection, and dose. The speed of the injection has a slight effect on the block, with slow injections resulting in a faster and more predictable block. Pregnant women require less local anaesthetic to achieve the same level of block as non-pregnant women, because of compression of the dural sac by the engorged epidural venous plexus. When performing a saddle block the patient should remain sitting for at least 5 minutes after injection of hyperbaric 'heavy' local anaesthetic.

Complications

See box on *Epidural anaesthesia*.

Post-procedure checks

- Spinal anaesthesia denervates the sympathetic chain, causing hypotension and bradycardia depending on the block height. Appropriate fluids, vasopressors or inotropes are administered.
- High spinal blockade affects active exhalation due to paralysis of the abdominal and intercostal muscles.
- Inspiration is less affected, as the phrenic nerve is spared, although the intercostal muscles may be affected.
- The block should be assessed with light touch, pinprick or ice 10–15 minutes later.

References

1. Zaric D, Christiansen C, Pace NL, *et al*. Transient neurologic symptoms (TNS) following spinal anaesthesia with lidocaine versus other local anaesthetics: a systematic review of randomized, controlled trials. *Anesth Analg* 2005; **100**: 1811–16.

The head

The skull

1. The bones of the skull

The skull consists of 22 bones.
It is subdivided into the neurocranium and the facial skeleton.

i) Neurocranium

Consisting of the following single bones:
- Frontal.
- Occipital.
- Sphenoid.
- Ethmoid.

And the following paired bones:

- Parietal.
- Temporal.

ii) Facial skeleton

Consisting of the following single bones:
- Mandible.
- Vomer.

And the following paired bones:
- Lacrimal.
- Nasal.
- Inferior nasal conchae.
- Maxillae.
- Zygomatic.
- Palatine.

Structures   Openings and foramina

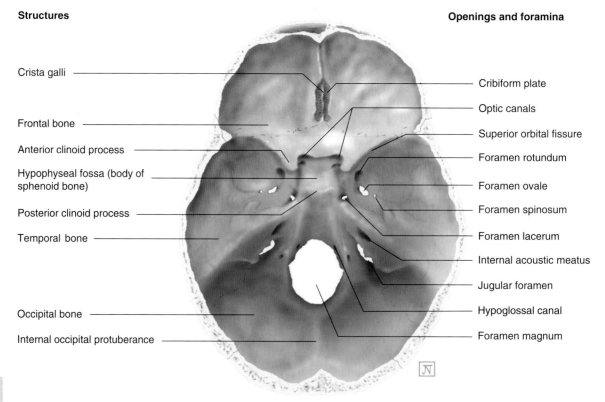

Crista galli

Frontal bone

Anterior clinoid process

Hypophyseal fossa (body of sphenoid bone)

Posterior clinoid process

Temporal bone

Occipital bone

Internal occipital protuberance

Cribiform plate

Optic canals

Superior orbital fissure

Foramen rotundum

Foramen ovale

Foramen spinosum

Foramen lacerum

Internal acoustic meatus

Jugular foramen

Hypoglossal canal

Foramen magnum

Figure 3.1 Superior view of the cranial base.

Structures		Openings and foramina

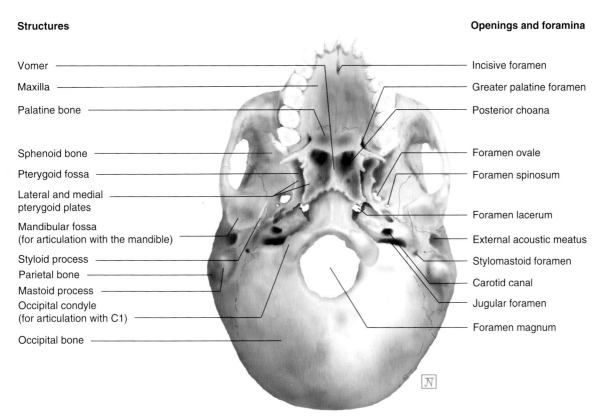

Vomer

Maxilla

Palatine bone

Sphenoid bone

Pterygoid fossa

Lateral and medial pterygoid plates

Mandibular fossa (for articulation with the mandible)

Styloid process

Parietal bone

Mastoid process

Occipital condyle (for articulation with C1)

Occipital bone

Incisive foramen

Greater palatine foramen

Posterior choana

Foramen ovale

Foramen spinosum

Foramen lacerum

External acoustic meatus

Stylomastoid foramen

Carotid canal

Jugular foramen

Foramen magnum

Figure 3.2 Inferior view of the cranial base.

2. The foramina of the skull

A summary of the structures passing through each foramen of the skull is given in Table 3.1.

Table 3.1 Structures passing through the foramina of the skull. CN, cranial nerve. Note: CN V_2 and the infraorbital vessels traverse the infraorbital fissure, but don't pass through it.

Foramina	Contents
Foramina in cribiform plate	Axons of olfactory nerve
Anterior and posterior ethmoidal formainae	Vessels and nerves of the same name
Optic canal	CN II and ophthalmic arteries
Superior orbital fissure	CN III, IV, branches of V_1 (frontal lacrimal, nasociliary), VI, and sympathetic fibres. Superior and inferior ophthalmic veins.
Inferior orbital fissure	Zygomatic branch of CNV_2, pterygopalatine branches, emissary vein.
Foramen rotundum	CN V_2
Foramen ovale	CN V_3 and accessory meningeal artery
Foramen spinosum	Meningeal branch of CN V_3. Middle meningeal artery and vein.
Foramen lacerum	Internal carotid artery, sympathetic and venous plexuses (pass across)
Jugular foramen	CN IX, X, XI. Superior bulb of internal jugular vein.
Foramen magnum	Medulla, meninges. Spinal roots of CN XI. Vertebral and spinal arteries. Dural veins.
Stylomastoid foramen	CN VII
Hypoglossal canal	CN XII

The brain

1. Components of the brain

i) The cerebrum

Consists of:
- The cerebral hemispheres – occupy the anterior and middle cranial fossae and extend over the cerebellum. Responsible for higher functions, initiation of motor commands and receipt of sensory information.
- The diencephalon – composed of the epithalamus, thalamus and hypothalamus. Responsible for consciousness, control of basal metabolic functions and hormonal control of the pituitary.

ii) The cerebellum

- Lies in the posterior cranial fossa, posterior to the pons and medulla.
- Consists of two lateral hemispheres separated by the vermis.
- Responsible for fine control and coordination of motor function and speech.

iii) The brainstem

Contains the nuclei of the cranial nerves.
Consists of:
- The midbrain.
- The pons.
- The medulla oblongata – contains the cardiorespiratory centres and is continuous with the spinal cord.

2. Ventricles of the brain

Cerebrospinal fluid (CSF) is formed at a rate of 500 ml/day by modified ependymal cells in the choroid plexuses found in the ventricular system of the brain. Passage of CSF through the ventricular system:
- From the two lateral ventricles (found in each of the cerebral hemispheres), CSF passes via an interventricular foramen into the third ventricle (surrounded by the diencephalon).
- CSF then flows via the cerebral aqueduct (of Sylvius, in the midbrain) to the fourth ventricle (surrounded by the pons and medulla).
- The fourth ventricle communicates with the central canal of the spinal cord.
- From the fourth ventricle, CSF drains to the subarachnoid space is via a median aperture (foramen of Magendie) and paired lateral apertures (foramina of Luschka).
- The subarachnoid space surrounds the brain and is in direct communication with the space surrounding the spinal cord.

CSF is resorbed by arachnoid granulations, where it then enters the venous system.
The total volume of CSF is 120 ml; this is exchanged approximately four times per day.

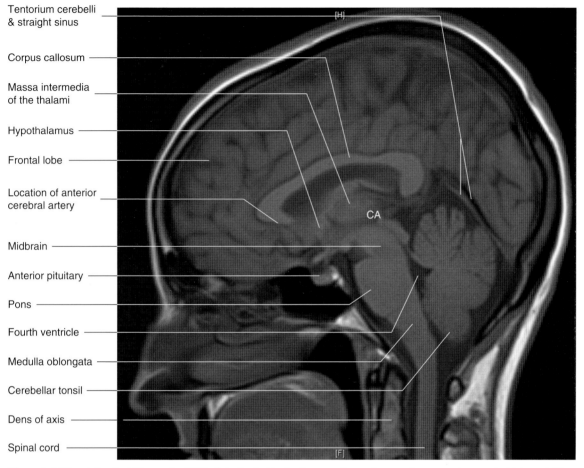

Tentorium cerebelli & straight sinus

Corpus callosum

Massa intermedia of the thalami

Hypothalamus

Frontal lobe

Location of anterior cerebral artery

Midbrain

Anterior pituitary

Pons

Fourth ventricle

Medulla oblongata

Cerebellar tonsil

Dens of axis

Spinal cord

CA

Figure 3.3 **T1-weighted midline sagittal MRI of the brain.** The basilar artery lies on the pons (but cannot be seen in this image). CA, cerebral aqueduct.

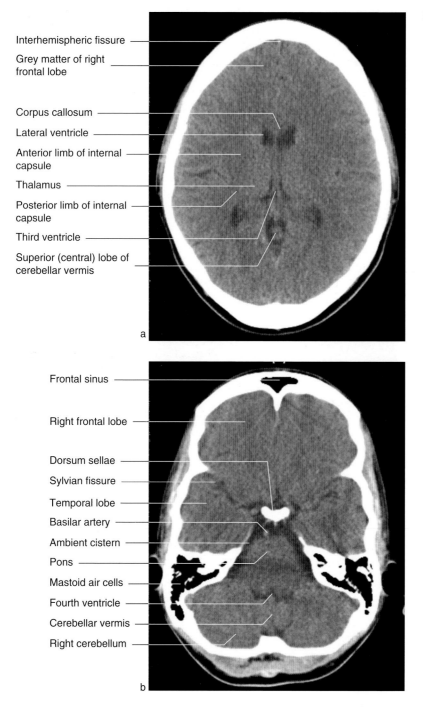

Interhemispheric fissure

Grey matter of right frontal lobe

Corpus callosum

Lateral ventricle

Anterior limb of internal capsule

Thalamus

Posterior limb of internal capsule

Third ventricle

Superior (central) lobe of cerebellar vermis

a

Frontal sinus

Right frontal lobe

Dorsum sellae

Sylvian fissure

Temporal lobe

Basilar artery

Ambient cistern

Pons

Mastoid air cells

Fourth ventricle

Cerebellar vermis

Right cerebellum

b

Figure 3.4 Normal transverse CT images of the brain.

3. Cranial meninges

The meninges are the dura mater, arachnoid mater and pia mater (from outermost to innermost). They surround, support and protect the brain and are in continuation with the meninges surrounding the spinal cord (see Chapter 2, *The spinal cord and meninges*, section 6).

i) Dura mater

A double-layered tough membrane.

The outer (periosteal) layer is the periosteal lining of the skull; the inner (meningeal) layer continues through the foramen magnum to surround the spinal cord.

The two layers separate to form dural infoldings:

- Falx cerebri – in the longitudinal fissure between the left and right cerebral hemispheres.
- Falx cerebelli – found between the cerebellar hemispheres.
- Tentorium cerebelli – separates the occipital cerebral hemispheres from the cerebellum. Attaches to the falx cerebri.
- Diaphragma sellae – covers the pituitary gland.

Dural venous sinuses form between the two layers of dura: e.g. the superior and inferior sagittal sinuses are found in the falx cerebri. They ultimately convey venous blood from the surface of the brain into the internal jugular vein.

ii) Arachnoid mater

A thin avascular layer which is held against the inner layer of dura by the pressure of CSF beneath it.

iii) Pia mater

A highly vascular layer which adheres to the brain and follows all of its contours.

Coning involves herniation of cerebral structures, either across dural infoldings or through the foramen magnum, due to increased intracranial pressure. This may be:

- Supratentorial – most commonly involves herniation under the falx cerebri (subfalcine herniation) resulting in abnormal posturing or coma. Alternatively it may involve compression against the tentorial opening. This may compress the diencephalon (central herniation – early fall in conscious level and abnormal posturing, progressing to medullary compression), or the medial uncus (uncal herniation – early third nerve compression and late fall in consciousness).
- Infratentorial – involving herniation of the cerebellum, which may be upward through the tentorium (midbrain compression and hydrocephalus), or downward through the foramen magnum (medullary compression).

4. Meningeal spaces

i) Extradural or epidural space

A potential space between the cranial bones and the outer periosteal layer of dura (the two are attached).

Becomes an actual space usually when arterial blood separates the two, e.g. following a torn meningeal vessel.

ii) Subdural space

A potential space between the dura and the arachnoid layers.

Becomes an actual space acutely when a bridging vessel (between the brain and a venous sinus) tears; hence bleeds are usually venous in nature.

Chronic subdural bleeds may arise due to rupture of fragile vessels in a neomembrane, which form between the dural and arachnoid layers.

iii) Subarachnoid space

An actual space that contains CSF, arteries and veins.

The circle of Willis is found within this space; subarachnoid bleeds are therefore usually arterial in nature.

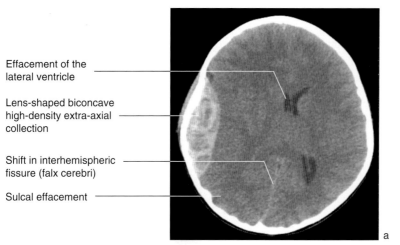

Effacement of the lateral ventricle

Lens-shaped biconcave high-density extra-axial collection

Shift in interhemispheric fissure (falx cerebri)

Sulcal effacement

a

Figure 3.5 Intracranial bleeds.
(a) Extradural bleed with mass effect.
(b) Acute subdural bleed with mass effect. (c) Subarachnoid bleed without ventricular blood or hydrocephalus.

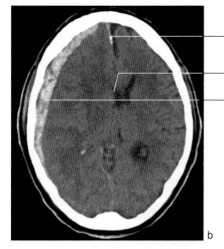

Shift in interhemispheric fissure (falx cerebri)

Effacement of the lateral ventricle

High-density crescentic extra-axial collection with a concave inner margin and a convex outer margin. It traverses the skull.

b

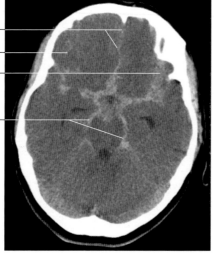

Parafalcine blood

Sulcal blood

Blood in Sylvian fissure

Blood in the basal cisterns

c

Major branches

Minor branches

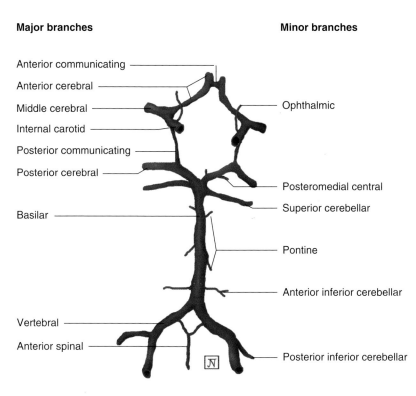

Anterior communicating

Anterior cerebral

Middle cerebral

Internal carotid

Posterior communicating

Posterior cerebral

Basilar

Vertebral

Anterior spinal

Ophthalmic

Posteromedial central

Superior cerebellar

Pontine

Anterior inferior cerebellar

Posterior inferior cerebellar

Figure 3.6 The arterial circle of Willis viewed from below (so that the internal carotid arteries are coming out of the page towards the reader). The vertebral arteries originate from the subclavian arteries. The circle of Willis is responsible for the blood supply to the brain.

5. Vascular supply

i) Arterial

Derived from:

- The left and right internal carotid arteries. Arise from the common carotid arteries in the neck. Enter the cranial cavity through the carotid canals in the temporal bone before passing across the foramen lacerum. Terminate by dividing into the anterior and middle cerebral arteries to supply the anterior circulation of the brain.
- The left and right vertebral arteries. Arise from the subclavian artery. Ascend through the foramina of the transverse processes of the cervical vertebrae before passing through the foramen magnum. Unite at the pons to form the basilar artery and supply the posterior circulation of the brain.

The anterior and posterior circulations are united by the cerebral arterial circle (of Willis) at the base of the brain (Figure 3.6).

The circle is found within the subarachnoid space. This anastomosis and the branches from it are responsible for supplying the entire brain with arterial blood.

Individual variation means that the true circle is only found in approximately one-third of the population.

ii) Venous

Blood from superficial and deep veins enters the dural venous sinuses (see section 4, *Cranial meninges, dura mater*, above).

The sinuses drain to the internal jugular veins, which leave the skull via the jugular foramen.

The cranial nerves

Cranial nerves are the peripheral nerves of the brain (in the same way that the spinal nerves are the peripheral nerves of the spine).

There are 12 pairs of cranial nerves, all of which originate from cranial nerve nuclei situated in the pons and the medulla.

Cranial nerves (CN) I and II are atypical.

Cranial nerves III, VII, IX and X carry parasympathetic preganglionic fibres to respective ganglia (see Chapter 1, section 7).

In the following text, nerves with predominantly motor function are described from the brain, progressing distally; those with predominantly sensory function are

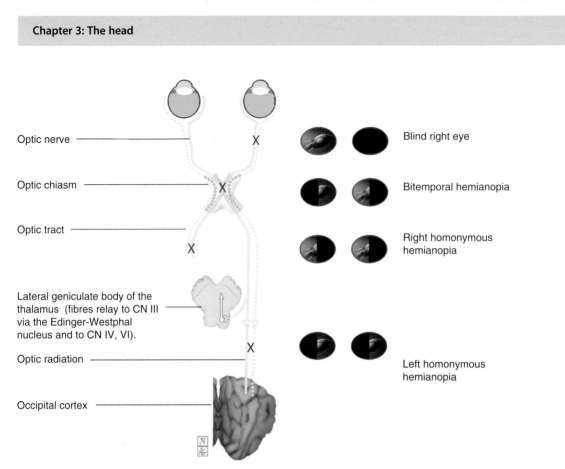

Optic nerve

Optic chiasm

Optic tract

Lateral geniculate body of the thalamus (fibres relay to CN III via the Edinger-Westphal nucleus and to CN IV, VI).

Optic radiation

Occipital cortex

Blind right eye

Bitemporal hemianopia

Right homonymous hemianopia

Left homonymous hemianopia

Figure 3.7 Schematic image of the optical pathway. Red crosses indicate possible sites of lesions within the pathway, and the subsequent visual disturbances are shown on the right of the image. (For clarity the pathway is only shown in full on the right-hand side of the image). Note how the nasal fibres decussate at the optic chiasm; the temporal fibres do not.

described from the periphery, moving proximally towards the brain.

1. Olfactory nerve (CN I)

- Special sense – smell.

Fibres originate in the olfactory mucosa (roof of nasal cavity, nasal septum and superior concha).
Pass through the cribiform plate of the ethmoid bone and enter the olfactory bulb.
Synapse with mitral cells to form the olfactory tract, passing to the temporal lobe.

2. Optic nerve (CN II)

- Special sense – vision.

Axons of retinal ganglion cells form the optic nerve.
The optic nerve passes through the optic canal to enter the middle cranial fossa, where the optic chiasm is formed.
Here nasal fibres (which serve the temporal visual fields) decussate; temporal fibres (which serve the nasal visual fields) do not.

Fibres continue in the optic tract to the lateral geniculate body (in the thalamus).
Some fibres then pass to the superior colliculus (ocular and pupillary reflexes via CN III (and hence the light reflex via the Edinger–Westphal nucleus), CN IV and VI); the remainder enter the optic radiation to pass to the occipital cortex.
Note: the meninges and subarachnoid space enclose the optic nerve. The meninges fuse with the sclera, explaining why raised CSF pressure creates papilloedema and how a total spinal can result from a misplaced eye block.

3. Oculomotor nerve (CN III)

- Motor to all external eye muscles except superior oblique and lateral rectus. Also supplies levator palpebrae superioris.
- Parasympathetic to sphincter pupillae (constriction) and ciliary muscles (accommodation).

Origin – somatic and visceral nuclei of CN III. The latter is known as the Edinger–Westphal nucleus and gives off preganglionic parasympathetic fibres, thereby mediating the light reflex.

Runs across the middle cranial fossa close to the posterior communicating artery.

Enters the cavernous sinus.

Passes through the superior orbital fissure and divides:

i) Superior division

Fibres to superior rectus, levator palpebrae superioris.

ii) Inferior division

Fibres to inferior and medial recti and inferior oblique.

Parasympathetic preganglionic fibres pass to the ciliary ganglion, where postsynaptic fibres (short ciliary nerves) then innervate sphincter pupillae and the ciliary muscles.

Raised intracranial pressure may compress CN III over the tentorium cerebelli, impeding parasympathetic supply to the eye, resulting in a pupil which is 'fixed and dilated'.

4. Trochlear nerve (CN IV)

- Motor to superior oblique muscle.

Has the longest intracranial course.

Passes through the cavernous sinus.

Passes through the superior orbital fissure to enter the orbit.

5. Trigeminal nerve (CN V)

- Sensory to the face, orbit, mucosa of nasal cavity and mouth, anterior two-thirds of the tongue, nasal and maxillary sinuses.
- Motor to the muscles of mastication, mylohyoid, digastric, tensor veli palatini, tensor tympani.

It has three major branches (Figure 3.8):

i) Ophthalmic nerve (CN V$_1$)

- Sensory only.
  Passes through the superior orbital fissure.

ii) Maxillary nerve (CN V$_2$)

- Sensory only.
  Passes through the foramen rotundum.

iii) Mandibular nerve (CN V$_3$)

- Sensory and motor. Note: supplies general sensation, not taste, to the anterior two-thirds of the tongue.
  Passes through the foramen ovale.

The sensory nerves converge on the trigeminal ganglion, found in a dural cavity known as Meckel's cave, 1 cm from the pons.

The trigeminal ganglion contains the cell bodies of the sensory nerves, so is equivalent to the dorsal root ganglion of a spinal nerve.

The motor fibres pass beneath the ganglion.

All nerves then pass to the trigeminal nerve nuclei in the brainstem. There are four nuclei: one motor and three sensory, corresponding to the final divisions of the nerve.

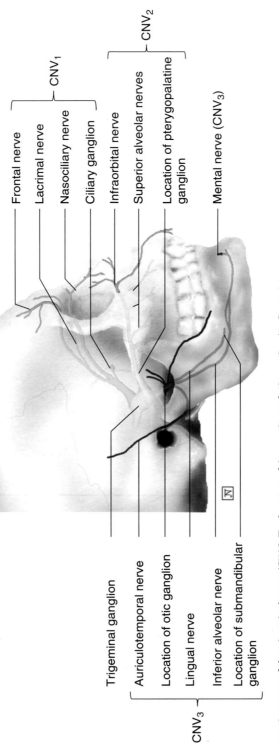

Frontal nerve

Lacrimal nerve

CNV_1

Nasociliary nerve

Ciliary ganglion

Infraorbital nerve

Superior alveolar nerves

CNV_2

Location of pterygopalatine ganglion

Mental nerve (CNV_3)

Trigeminal ganglion

Auriculotemporal nerve

Location of otic ganglion

Lingual nerve

Inferior alveolar nerve

Location of submandibular ganglion

CNV_3

Figure 3.8 Anatomy of the trigeminal nerve (CN V). The functions of the ganglia are further outlined in Figure 1.4.

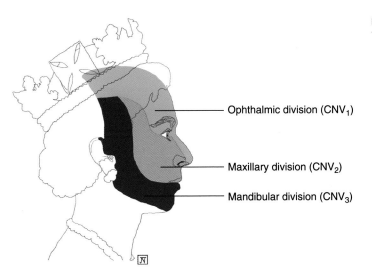

Figure 3.9 Cutaneous distribution of the trigeminal (CN V) nerve.

Ophthalmic division (CNV$_1$)

Maxillary division (CNV$_2$)

Mandibular division (CNV$_3$)

Trigeminal nerve block

Introduction

Trigeminal neuralgia is a rare condition that produces severe neuropathic facial pain in one or more divisions of the trigeminal nerve. The symptoms are treated with oral neuropathic drugs such as carbamazepine[1] once intracranial pathology has been excluded by neurological examination and imaging. Patients who do not respond to pharmacological therapy may have an initial local anaesthetic block of the trigeminal ganglion as a diagnostic test, followed by chemical neurolysis or radiofrequency lesioning.[2]

Indications

Trigeminal neuralgia.

Technique

X-ray image intensifier.

Landmark

The patient is in the supine position with chin up, neck extended and head rotated 30° away from the affected side. The axis of the image intensifier c-arm is aligned with the foramen ovale by using 30° of caudal angulation of the beam. Local anaesthetic is infiltrated 2–3 cm lateral to the corner of the mouth and a 22G 10 cm spinal needle is advanced under x-ray control towards the foramen ovale. Lateral screening may be required to check the needle depth. The patient may experience paraesthesia in one of the divisions of the trigeminal nerve when the needle enters the foramen ovale, usually at a depth of 1–1.5 cm. Correct positioning is likely if CSF is seen to leak from the needle or injected contrast shows filling of the cistern formed by extension of arachnoid mater into Meckel's cave. The needle is gradually withdrawn until CSF is no longer aspirated. 1 ml of local anaesthetic produces dense analgesia.

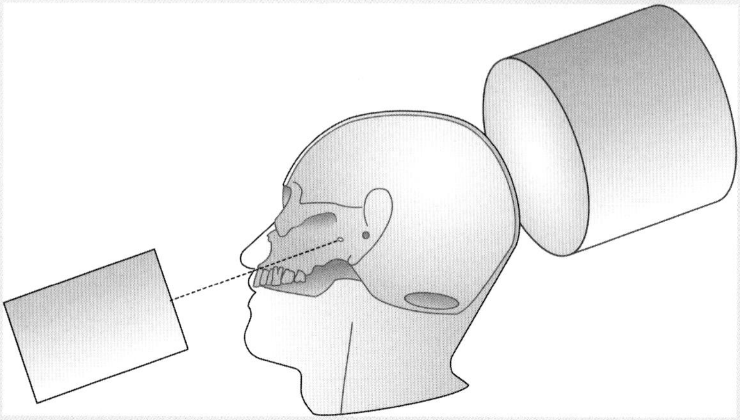

Figure 3.10 X-ray screening for a trigeminal nerve block.

Complications

After local anaesthetic:
- Generalised seizures, following carotid artery injection.
- Persistent paraesthesia.
- Total spinal.
- Haematoma of the cheek.

After neurolysis:
- Facial numbness.
- Reduced or abolished corneal reflex.
- Weakened muscles of mastication.
- Rarely paralysis of third or fourth cranial nerves.

Post-procedure checks

- The patient is sat upright and observed for at least 1 hour following the procedure.

References

1. Sidebottom A, Maxwell A. The medical and surgical management of trigeminal neuralgia. *J Clin Pharm Ther* 1995; **20**: 31–5.
2. Kondziolka D, Lunsford L, Young R, *et al.* Stereotactic radiosurgery for trigeminal neuralgia: A multiinstitutional study using the gamma unit. *J Neurosurg* 1996; **84**: 940–5.
3. Van Kleef M, Van Genderen W, Narouze S, *et al.* 1. Trigeminal neuralgia. *Pain Pract* 2009; **9**: 252–9.

6. Abducent nerve (CN VI)

- Motor to lateral rectus.

Bends sharply over the petrous temporal bone (where it may become compressed).
Enters the cavernous sinus.
Passes through the superior orbital fissure to enter the orbit.

7. Facial nerve (CN VII)

- Motor to muscles of facial expression, stapedius (of middle ear), stylohyoid, digastric.
- Sensation from skin of external auditory meatus.
- Special sense – taste from anterior two-thirds of tongue, floor of mouth, palate.

- Parasympathetic to submandibular and sublingual salivary glands (via the submandibular ganglion), lacrimal glands (via the pterygopalatine ganglion), glands of nasal cavity and palate.

Passes with CN VIII into the internal auditory meatus before passing through the facial canal in the temporal bone, where the following nerves arise:

i) The greater petrosal nerve

Preganglionic fibres to the pterygopalatine ganglion. Postganglionic fibres are secretomotor to the lacrimal gland.

ii) Nerve to stapedius

iii) Chorda tympani nerve

Taste to the anterior two-thirds of tongue and soft palate.
Also provides preganglionic fibres to the submandibular ganglion.
Postganglionic fibres are secretomotor to submandibular and sublingual glands.

CN VII then exits the skull via the stylomastoid foramen. Enters the parotid gland and divides into six terminal branches which supply the muscles of facial expression (posterior auricular, temporal, zygomatic, buccal, mandibular, cervical. Mnemonic – Perhaps The Zoo Breeds Meer Cats).

8. Vestibulocochlear nerve (CN VIII)

- Special sense – hearing and balance.

Composed of two nerves:

i) The vestibular nerve

Innervates the utricle, saccule and semicircular ducts. Concerned with balance.

ii) The cochlear nerve

Innervates the spiral organ (of Corti). Concerned with hearing.

Both nerves unite in the internal auditory meatus before entering the brainstem.

9. Glossopharyngeal nerve (CN IX)

- Sensory from mucosa of pharynx, posterior third of tongue, middle ear, Eustachian tube, carotid sinus and carotid body.

- Special sense – taste from posterior third of the tongue.
- Motor to stylopharyngeus.
- Parasympathetic to parotid gland and glands in the posterior third of the tongue.

Derived from four cranial nerve nuclei in the medulla. Several rootlets leave the medulla and coalesce to exit the skull via the jugular foramen.
It then passes between the superior and middle constrictors of the pharynx where it breaks up into terminal branches:

i) Tympanic branch

Supplies the tympanic cavity.
Continues as the lesser petrosal nerve which carries preganglionic parasympathetic fibres to the otic ganglion.
Postganglionic fibres are secretomotor to the parotid gland.

ii) Carotid nerve

Supplies the carotid sinus and body.

iii) Terminal branches

As indicated above.

10. Vagus nerve (CN X)

- Sensory from the inferior pharynx and larynx, thoracic and abdominal viscera.
- Special sense – taste from the epiglottis.
- Motor to soft palate, pharynx, intrinsic laryngeal muscles and palatoglossus.
- Parasympathetic to thoracic and abdominal viscera as far distally as the splenic flexure.

Originates from three nuclei in the medulla:
- The dorsal nucleus of the vagus (parasympathetic).
- The nucleus ambiguus (motor).
- The nucleus of the solitary tract (sensory).

Nine to ten rootlets emerge from the medulla and coalesce before leaving the skull through the jugular foramen.
Both vagal trunks then descend within the carotid sheath between the internal jugular vein and the internal and (more distally) common carotid arteries.
The paths of the left and right vagus nerves differ. The salient points of reference are:

i) Right vagus

Passes anterior to the right subclavian artery as it gives off the right recurrent laryngeal nerve.

Passes posterior to the right brachiocephalic vein.

ii) Left vagus

Passes between the left subclavian and left common carotid arteries.

Passes over the arch of the aorta, giving off the left recurrent laryngeal nerve.

Both nerves pass posterior to the root of their respective lungs and then pass into the abdomen through the oesophageal hiatus (at T10).

iii) Significant branches

- Jugular foramen – meningeal, auricular.
- Neck – pharyngeal, superior laryngeal, right recurrent laryngeal, superior cardiac.
- Thorax – inferior cardiac, left recurrent laryngeal, branches to the pulmonary and oesophageal plexuses.
- Abdomen – gastric, hepatic, intestinal (as far as the splenic flexure), branches to the coeliac plexuses.

11. Accessory nerve (CN XI)

- Motor to sternocleidomastoid (SCM) and trapezius muscles.

Previously thought to have spinal and cranial rootlets. The cranial rootlets are now viewed as part of the vagus nerve.

Spinal rootlets emerge from the spinal nucleus, a column of ventral horn cells stretching from C1 to C5 of the spinal cord.

The rootlets pass superiorly through the foramen magnum, then pass out of the skull through the jugular foramen.

The nerve then descends along the internal carotid artery, innervating SCM before passing across the posterior triangle of the neck to innervate trapezius.

12. Hypoglossal nerve (CN XII)

- Motor to the muscles of the tongue (with the exception of palatoglossus).

Small rootlets emerge from the medulla, fuse and exit the skull via the hypoglossal canal.

Passes around the internal carotid artery then moves superiorly to innervate the tongue.

Conveys some fibres from C1 and C2 to their final destinations.

13. Cranial nerve summary

The cranial nerves are summarised in Table 3.2.

Table 3.2 Summary of the cranial nerves.

Cranial nerve	Components	Action	Foramina
Olfactory (I)	Special sense	Smell	Cribiform plate of ethmoid
Optic (II)	Special sense	Vision	Optic canal
Oculomotor (III)	Motor	All external eye muscles (except superior oblique and lateral recuts) Also levator palpebrae superioris	Superior orbital fissure
	Parasympathetic	Sphincter pupillae and ciliary muscles	
Trochlear (IV)	Motor	Superior oblique	Superior orbital fissure
Trigeminal (V):			
CN V$_1$	Sensory	Upper face, cornea, nasal cavity	Superior orbital fissure
CN V$_2$	Sensory	Mid-face, maxillary teeth & sinuses	Foramen rotundum
CN V$_3$	Sensory and motor	Lower face & mandibular teeth, general sensation – anterior 2/3 of tongue Muscles of mastication, mylohyoid, digastric, tensor veli palatini, tensor tympani	Foramen ovale
Abducent (VI)	Motor	Lateral rectus	Superior orbital fissure
Facial (VII)	Motor	Muscles of facial expression, stapedius, stylohyoid, digastric	Internal auditory meatus then stylomastoid foramen
	Sensation	External auditory meatus	
	Special sense	Taste – anterior 2/3 of tongue, floor of mouth, palate	
	Parasympathetic	Submandibular, sublingual, lacrimal glands	
Vestibulocochlear (VIII)	Special sense	Hearing and balance	Internal auditory meatus

Table 3.2 (cont.)

Cranial nerve	Components	Action	Foramina
Glossopharyngeal (IX)	Sensory	Mucosa of pharynx, posterior 1/3 of tongue, middle ear, Eustachian tube, carotid sinus and carotid body	Jugular foramen
	Special sense	Taste from posterior 1/3 of tongue	
	Motor	Stylopharyngeus	
	Parasympathetic	Parotid gland, posterior 1/3 tongue	
Vagus (X)	Sensory	Inferior pharynx, larynx, thoracic and abdominal viscera	Jugular foramen
	Special sense	Taste from epiglottis	
	Motor	Soft palate, pharynx, intrinsic laryngeal muscles and palatoglossus	
	Parasympathetic	Thoracic and abdominal viscera as far distally as the splenic flexure	
Accessory (XI)	Motor	Sternocleidomastoid and trapezius	Foramen magnum then jugular foramen.
Hypoglossal (XII)	Motor	Muscles of the tongue (except palatoglossus)	Hypoglossal canal

Brainstem death testing

As the cranial nerve nuclei are all located in the brainstem, the series of tests involved in brainstem death testing aims to challenge the cranial nerves and therefore assess the function of the brainstem. The following applies to children over 2 months of age and adults.

Readers are referred to the full code of practice from the Academy of Medical Royal Colleges prior to engaging in these tests.

1. Aetiology

The aetiology of the patient's condition should be known. This should be, without question, the irreversible cause for the coma.

2. Diagnosis

Requires the following:

i) Necessary preconditions

Apnoeic coma requiring positive-pressure ventilation.
Irremediable structural brain damage caused by a disorder which can lead to brain death.

ii) Necessary exclusions

Primary hypothermia (target: core temperature > 34 °C).
Primary circulatory, metabolic or endocrine disorder (targets: MAP > 60 mmHg, $PaCO_2$ < 6.0 kPa, PO_2 10 kPa, pH 7.35–7.45, Na 115–160 mmol/L, K > 2.0 mmol/L, Mg and PO_4 0.5–3.0 mmol/L, glucose 3.0–20.0 mmol/L).

Drug intoxication or sedatives.
Paralysing drug or condition.
Abnormal posturing.

iii) Clinical findings

Absent cranial nerve reflexes:

- Pupillary light (CN II – afferent; CN III – efferent).
- Corneal (CN V – afferent; CN VII – efferent).
- Oculovestibular (CN VIII – afferent; CN III and VI – efferent).
- Gag (CN IX – afferent; CN X – efferent).
- Cough (CN X – afferent. The efferent pathway involves CN X, the phrenic nerve and multiple intercostal nerves).
- Motor response in a cranial nerve distribution to any peripheral stimulus (CN V, or peripheral nerves – afferent; CN VII – efferent).
- No respiratory effort despite $PaCO_2$ > 6.0 kPa (> 6.5 kPa if chronic CO_2 retention) and pH < 7.40 with adequate oxygenation (CN IX and central chemoreceptors – afferent; phrenic and intercostal nerves – efferent).

3. Repetition of tests

The tests should be performed by at least two medical practitioners who have been registered for at least 5 years and are competent in the conduct and interpretation of brainstem death testing.

Testing should be carried out by the doctors acting together, but should be performed on two separate occasions.

The legal time of death is when the first set of tests yields no brainstem function.

The eye and orbit

1. The orbit

Surrounds and protects the eye and its associated muscles, vessels and nerves.

Roughly conical in shape, with the apex directed posteriorly and slightly medially towards the optic canal.

Approximate orbital volume is 30 ml; 7 ml is occupied by the globe and muscle cone and 23 ml is occupied by loose connective tissue.

i) Walls of the orbit

The orbital margins and walls are illustrated in Figure 3.11.

The medial walls, either side of the nasal cavity, are parallel to one another.
The lateral walls are at right angles to one another.
The medial and lateral walls are each approximately 50 mm in length.

ii) Orbital contents

- The eyeball.
- The orbital muscles.
- The orbital vessels.
- The optic and orbital nerves.
- The lacrimal apparatus.
- Fat and connective tissue.

2. The eyeball

Sits high and lateral within the orbit (Figure 3.12).

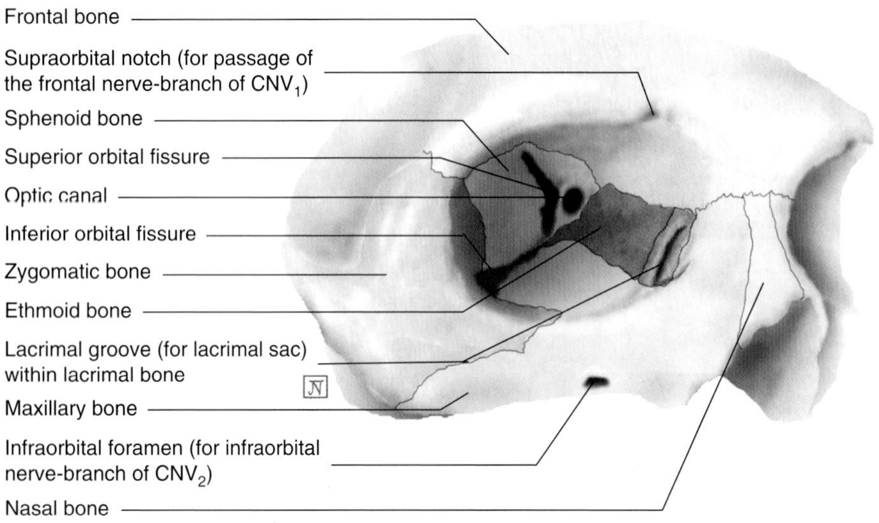

Frontal bone

Supraorbital notch (for passage of the frontal nerve-branch of CNV_1)

Sphenoid bone

Superior orbital fissure

Optic canal

Inferior orbital fissure

Zygomatic bone

Ethmoid bone

Lacrimal groove (for lacrimal sac) within lacrimal bone

Maxillary bone

Infraorbital foramen (for infraorbital nerve-branch of CNV_2)

Nasal bone

Figure 3.11 The bones of the orbit.

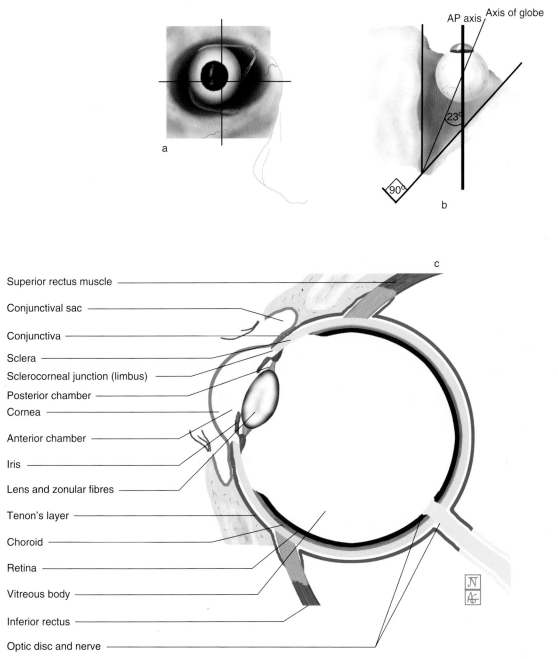

a

AP axis / Axis of globe

23°

90°

b

c

Superior rectus muscle

Conjunctival sac

Conjunctiva

Sclera

Sclerocorneal junction (limbus)

Posterior chamber

Cornea

Anterior chamber

Iris

Lens and zonular fibres

Tenon's layer

Choroid

Retina

Vitreous body

Inferior rectus

Optic disc and nerve

Figure 3.12 The anatomy of the eyeball. Note (a) how the globe sits high and lateral within the orbit and (b) lies at an axis of 23° to the AP axis. (c) The anatomical contents of the globe. Note how the lens is suspended from the ciliary body (unlabelled) via zonular fibres. The Tenon's fascia merges with the dura posteriorly, and thus the optic nerve becomes surrounded by CSF posteriorly. Adapted from Parness G, Underhill S. Regional anaesthesia for intraocular surgery. *Contin Educ Anaesth Crit Care Pain* 2005; **5**: 93–7.

The distance from the orbital rim to the optic foramen is 42–54 mm, with a typical axial length being 20–25 mm. Consists of:

i) Outer fibrous layer

Known as the sclera in its posterior five-sixths (this is the white of the eyeball). The sclera is continuous with the dural sheath of the optic nerve.
Known as the cornea in its anterior one-sixth (this is transparent).
The sclera and the cornea are continuous with one another.
Covered by the conjunctiva, which is reflected onto the internal surface of the eyelids.

ii) Middle vascular layer

Consists of the choroid, ciliary body and iris.
The choroid is attached firmly to the retina, to which it provides the vascular supply. It terminates anteriorly in the ciliary body.
Ciliary processes in the ciliary body secrete aqueous humour, which fills the anterior and posterior chambers and provides nutrients to the avascular cornea and lens.
The aqueous humour is drained via the canal of Schlemm at the iridocorneal angle to venous plexuses.
The size of the pupil is controlled by the sphincter and dilator pupillae muscles, under parasympathetic and sympathetic control respectively.

iii) Inner retinal layer

The retina is divided into an outer pigment layer and an inner neural layer.
The optic disc (which is insensitive to light) represents the site of exit of the optic nerve.

3. The bulbar or Tenon's fascia

A thin fascial sheath surrounding the eyeball.
Found between the sclera and the conjunctiva.
Attached posteriorly to the sclera, close to the optic nerve (where it fuses with the dura surrounding the nerve) and anteriorly to the sclerocorneal junction. It therefore does not cover the cornea.
Separates the eyeball from the other orbital contents; the potential space formed between the eyeball and the Tenon's fascia (the episcleral or sub-Tenon's space) allows free movement of the eyeball.
The tendons which insert into the eyeball pierce the Tenon's fascia. The fascia then reflects back onto the muscles.
Local anaesthetic deposited within the sub-Tenon's space (see box on *Sub-Tenon's block*, below) acts on the sensory and motor nerves which pass through the space prior to entering the globe. Diffusion of anaesthetic agent towards the optic nerve may affect vision.

Sub-Tenon's block

Introduction

Sub-Tenon's block provides anaesthesia for eye surgery including cataract and retinal operations. Local anaesthetic is injected into the episcleral space.

Indications

Cataract and retinal surgery.

Specific contraindications

- Previous scleral banding, detachment surgery and medial rectus surgery (relative contraindications).
- The risk of globe perforation is increased in myopic eyes, as the globe is longer with a thinner sclera and may have an irregular surface (staphylomata).

Pre-procedure checks

Eye surgery carries a low risk of perioperative morbidity and mortality, but older patients require careful preoperative assessment as they frequently have significant comorbidities.[1] Patients with a chronic cough, shortness of breath, severe reflux, inability to communicate or anxiety may be unable to lie still during the surgery, and therefore general anaesthetic may be required. Basic intraoperative monitoring is recommended, and intravenous access is required.

Technique

Landmark

Proxymetacaine 0.5% drops are instilled into the eye. A speculum is inserted to retract the eyelids. The conjunctiva is grasped in the inferior nasal quadrant 5–10 mm from the corneal limbus with Moorfield's forceps. Westcott spring scissors are used to make an opening in the conjunctiva and Tenon's capsule to gain access to the sub-Tenon's space. A blunt curved sub-Tenon's cannula is inserted into the sub-Tenon's space (Figure 3.13).

Low volume (3–5 ml) provides analgesia of the anterior segment for cataract surgery but only partial akinesia of the globe and lids. Volumes up to 11 ml have been used to promote spread to the extraocular muscle sheaths, thereby providing akinesia and allowing surgery of the posterior segment.

Complications

In a series of 6000 cases there were no serious complications.[2]
- Subconjuctival haematoma (7%).
- Subconjuctival spread of local anaesthetic (chemosis) is common after injection in the sub-Tenon's space (6%), and is treated by compression.
- 1 patient of 6000 cancelled because of haematoma in surgical field.

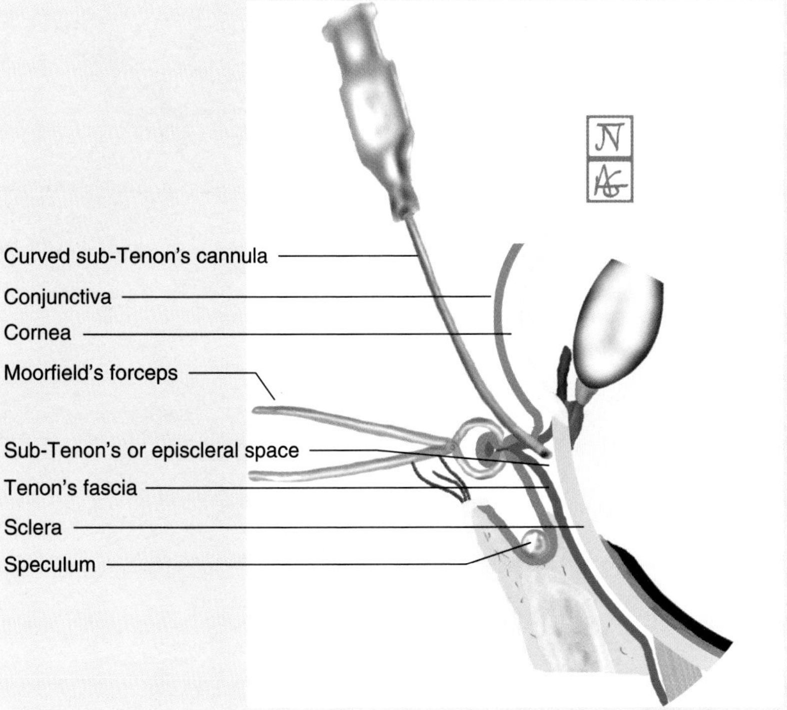

Curved sub-Tenon's cannula
Conjunctiva
Cornea
Moorfield's forceps

Sub-Tenon's or episcleral space
Tenon's fascia
Sclera
Speculum

Figure 3.13 Anatomical depiction of a sub-Tenon's block. Note that the Tenon's fascia terminates anteriorly by attaching to the sclerocorneal junction, so a successful injection must involve introducing the cannula 5–10 mm from the sclerocorneal junction (limbus).

References

1. Katz J, Feldman MA, Bass EB, *et al.* Adverse intraoperative medical events and their association with anesthesia management strategies in cataract surgery. *Ophthalmology* 2001; **108**: 1721–6.
2. Guise P. Sub-Tenon anesthesia: a prospective study of 6,000 blocks. *Anesthesiology* 2003; **98**: 964–8.

4. The orbital muscles

i) Levator palpebrae superioris

From the roof of the orbit to the skin of the upper eyelid.
Action: elevates eyelid.

ii) Rectus muscles

All originate from the common tendinous ring (which surrounds the optic canal) and insert into the globe, posterior to the sclerocorneal junction (Figure 3.14).
Together they form the muscle cone, which encloses the sensory and motor nerves, the ciliary ganglion, the optic nerve and the retinal artery and vein.
Action:
- Superior rectus – elevation.
- Inferior rectus – depression.
- Medial rectus – adduction.
- Lateral rectus – abduction.

iii) Oblique muscles

- Superior oblique:
From the sphenoid bone it passes through the trochlea before attaching superolaterally behind the equator.
Action – abduction, depression, medial rotation ('down and out').
- Inferior oblique:
From the maxilla it passes laterally, before attaching posterolaterally behind the equator.
Action – abduction, elevation, lateral rotation.

All muscles are innervated by CN III, with the exception of lateral rectus (innervation from CN VI) and superior oblique (innervation from CN IV) muscles.

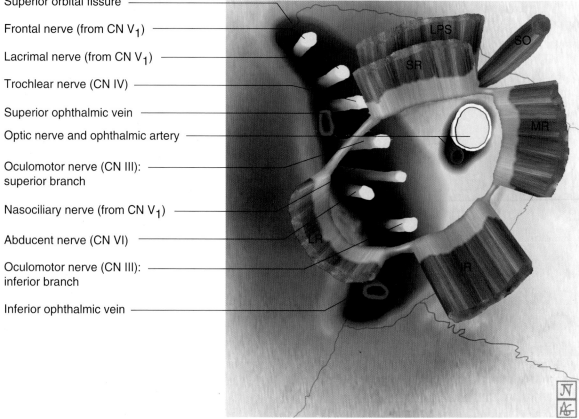

Superior orbital fissure

Frontal nerve (from CN V$_1$)

Lacrimal nerve (from CN V$_1$)

Trochlear nerve (CN IV)

Superior ophthalmic vein

Optic nerve and ophthalmic artery

Oculomotor nerve (CN III): superior branch

Nasociliary nerve (from CN V$_1$)

Abducent nerve (CN VI)

Oculomotor nerve (CN III): inferior branch

Inferior ophthalmic vein

Figure 3.14 Structures passing through the superior orbital fissure and optic canal. The inferior orbital fissure is not shown. The optic nerve and ophthalmic artery pass through the optic canal; all other nerves pass through the superior orbital fissure. Mnemonic – Fish Like To Occupy the North Atlantic (Frontal, Lacrimal, Trochlear, Oculomotor, Nasociliary, Abducent). Muscles: SR, superior rectus; LR, lateral rectus; IR, inferior rectus; MR, medial rectus; SO, superior oblique; LPS, levator palpebrae superioris. Note how all the recti originate from the common tendinous ring, and in doing so form the muscle cone.

In addition, levator palpebrae superioris has some sympathetic innervation (explaining the ptosis which results with Horner's syndrome).

5. Neurovascular supply

i) Arterial

Derived from the ophthalmic artery, a branch of the internal carotid artery.

The ophthalmic artery passes through the optic canal with the optic nerve and divides into several branches that supply the orbital contents (e.g. the central artery of the retina).

ii) Venous

Superior and inferior ophthalmic veins (the latter drains into the former), which pass through the superior orbital fissure to drain into the cavernous sinus (a potential route of ingress for tracking infection).

An emissary vein drains directly to the pterygoid venous plexus.

iii) Nervous

See Figure 3.14.
- Motor: CN III, IV and VI (see section 4, above).
- Special sensory: CN II (see section 4, above).
- Sensory:
  CN V_1 (the ophthalmic division of the trigeminal nerve) enters the orbit via the superior orbital fissure and divides into the lacrimal, frontal and nasociliary nerves (the nasociliary nerve gives off intraconally and supplies the upper eyelid extraconally.

  CN V_2 (the maxillary branch of the trigeminal nerve) exits the skull via the foramen rotundum, passes across the inferior orbital fissure and through the infraorbital canal and emerges as the infraorbital nerve at the infraorbital foramen (Figures 3.8 and 3.11). It supplies the lower lid, so delivers only extraconal fibres.
- Autonomic:
  Sympathetic – preganglionic fibres ascend in the sympathetic trunk to synapse with postganglionic fibres in the superior cervical ganglion. The postganglionic fibres ascend as a plexus with the internal carotid artery and pass through the superior orbital fissure to join the long ciliary nerves and innervate dilator pupillae (causing mydriasis).
  Parasympathetic – preganglionic fibres from CN III enter the orbit through the superior orbital fissure and synapse with postganglionic fibres at the ciliary ganglion, found within the muscle cone. The postganglionic fibres are the short ciliary nerves which innervate sphincter pupillae and the ciliary muscles (causing meiosis and accommodation).
  Postganglionic parasympathetic fibres also arise from the pterygopalatine ganglion (innervated by CN VII). They enter the orbit through the inferior orbital fissure and are secretomotor to the lacrimal gland.

Peribulbar block

Introduction

Peribulbar block provides anaesthesia for eye surgery by injection of local anaesthetic into the extraconal space. Anaesthetic spreads intraconally, as there is no membrane separating the extra- and intraconal compartments.[1]

Indications

Anaesthesia of the eye and orbit.

Specific contraindications

- The risk of globe perforation is increased in myopic eyes, as the globe is longer with a thinner sclera and may have an irregular surface (staphylomata).
- Anticoagulation (risk of retrobulbar haemorrhage; a sub-Tenon's approach is safer).

Pre-procedure checks

- Identify significant comorbidities and ability to lie still during surgery.[2]
- Monitoring and intravenous access is required.

Technique

25G 25mm needle.

Landmark

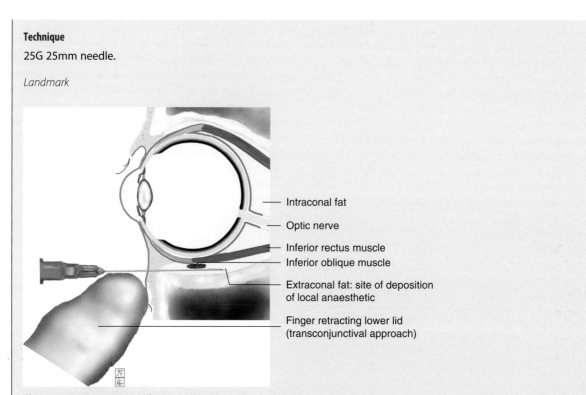

Intraconal fat

Optic nerve

Inferior rectus muscle
Inferior oblique muscle

Extraconal fat: site of deposition
of local anaesthetic

Finger retracting lower lid
(transconjunctival approach)

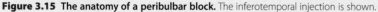

Figure 3.15 The anatomy of a peribulbar block. The inferotemporal injection is shown.

Proxymetacaine 0.5% drops are applied at the lateral and medial aspects of the eye. 10 ml of 2% lidocaine with or without adrenaline 1:200,000 is prepared. Hyaluronidase (15 units/ml) may be added to increase the spread of local anaesthetic but is associated with anaphylaxis and sterile abscess formation. The patient looks at a fixed point with the eyes in a neutral position. The injection may be transconjunctival (through a topically anaesthetised conjunctiva) or transcutaneous.

The lower lid is retracted manually and the needle (with the bevel facing superiorly) is introduced at a point inferior and temporal to the globe in line with the edge of the iris, or alternatively midway between this point and the lateral canthus. The needle is advanced parallel to the orbital floor and should not cause rotation of the globe. When the needle tip is past the equator the needle is redirected slightly upwards and inwards to avoid the bony orbital margin. The bevel of the needle is orientated towards the globe and advanced to a maximum depth of 25 mm. After negative aspiration, local anaesthetic is injected (6–12 ml).

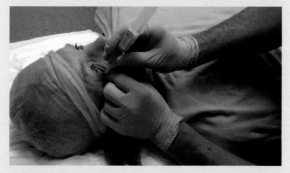

Figure 3.16 The anatomy of a peribulbar block – inferotemporal injection. A 25G 25 mm needle is used with the bevel facing superiorly. An idea as to the depth of the needle tip is therefore possible, given a knowledge of the axial length (available preoperatively – usually 22–25 mm). The site of injection should be posterior to the equator, as shown in Figure 3.15.

A second injection may be performed, to supplement the block, through the conjunctiva on the nasal side. The needle is inserted medial to the caruncle and directed vertically, parallel to the medial orbital wall. At a depth of 15 mm or less, and after negative aspiration, up to 5 ml of local anaesthetic is injected.

External compression with a Honan balloon (20–30 mmHg for 5–10 minutes) may be used to lower intraocular pressure.

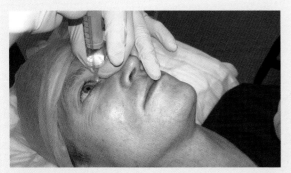

Figure 3.17 The anatomy of a peribulbar block – medial canthal injection.

Complications

The primary cause of serious complication is needle misplacement.

The eye:
- The risk of globe perforation and rupture is 1 in 350 to 7 in 50,000,[3] increasing with axial lengths greater than 26 mm.
- Arterial puncture may lead to retrobulbar haemorrhage and compressive haematoma, which may threaten retinal perfusion.
- Optic nerve damage is rare but may result in blindness.
- Chemosis from subconjunctival spread of local anaesthetic.

Systemic:
- Arterial injection, leading to seizures.[4]
- Injection under the dura of the optic nerve, leading to a total spinal.
- Oculocardiac reflex from traction on the globe (trigeminal afferent; vagal efferent), producing bradycardia.

Post-procedure checks

The signs of a successful block include:
- Ptosis.
- Akinesia.
- Inability to close the eye once opened.

References

1. Demediuk O, Dhaliwal R, Papworth D, *et al*. A comparison of peribulbar and retrobulbar anesthesia for vitreoretinal surgical procedures. *Arch Ophthalmol* 1995; **113**: 908–13.
2. Katz J, Feldman MA, Bass EB, *et al*. Adverse intraoperative medical events and their association with anesthesia management strategies in cataract surgery. *Ophthalmology* 2001; **108**: 1721–6.
3. Edge R, Navon S. Scleral perforation during retrobulbar and peribulbar anesthesia: Risk factor and outcome in 50,000 consecutive injections. *J Cataract Refract Surg* 1999; **25**: 1237–44.
4. Aldrete J, Romo-Salas F, Arora S, *et al*. Reverse arterial blood flow as a pathway for central nervous system toxic responses following injection of local anaesthetics. *Anesth Analg* 1978; **57**: 428–33.

Retrobulbar block

Introduction

Retrobulbar block provides anaesthesia of the eye and orbit by injection of local anaesthetic inside the intraconal space. The nerve supply to the superior oblique muscle is extraconal, and therefore total akinesia may not be achieved. Retrobulbar blocks are now rarely performed, as the needle trajectory is towards the optic nerve and close to the globe.

Indications

Anaesthesia of the eye and orbit.

Specific contraindications

- The risk of globe perforation is increased in myopic eyes, as the globe is longer with a thinner sclera and may have an irregular surface (staphylomata).
- Anticoagulation (risk of retrobulbar haemorrhage; a sub-Tenon's approach is safer).

Pre-procedure checks

- Identify significant comorbidities and ability to lie still during surgery.[1,2]
- Monitoring and intravenous access is required.

Technique

25G 25mm needle.

Landmark

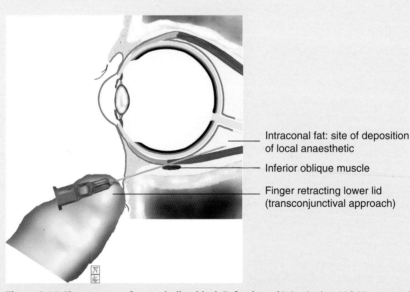

Intraconal fat: site of deposition of local anaesthetic

Inferior oblique muscle

Finger retracting lower lid (transconjunctival approach)

Figure 3.18 The anatomy of a retrobulbar block (inferolateral injection). A 25G 25 mm needle is used with the bevel facing superiorly. An idea as to the depth of the needle tip is therefore possible, given a knowledge of the axial length (available preoperatively – usually 22–25 mm). The maximum recommended depth of injection is 25–35 mm; this depth is therefore approximated when the needle hub is in line with the anterior surface of the globe.

Proxymetacaine 0.5% drops are applied at the lateral and medial aspects of the eye. 10 ml of 2% lidocaine with or without adrenaline 1:200,000 is prepared. Hyaluronidase (15 units/ml) may be added to increase the spread of local anaesthetic but is associated with anaphylaxis and sterile abscess formation. The patient looks at a fixed point with the eyes in a neutral position. The injection may be transconjunctival (through a topically anaesthetised conjunctiva) or transcutaneous.

The lower lid is retracted manually and the needle (with the bevel facing superiorly) is introduced at a point inferior and temporal to the globe in line with the edge of the iris, or alternatively midway between this point and the lateral canthus. The needle is advanced parallel to the orbital floor and should not cause rotation of the globe. When the needle tip is past the equator the needle is redirected upwards and inwards to enter the space behind the globe between the inferior and lateral rectus muscles to a maximum depth of 25–35 mm. After negative aspiration, 3–5 ml of local anaesthetic is injected slowly.

External compression with a Honan balloon (20–30 mmHg for 5–10 minutes) may be used to lower intraocular pressure.

Complications

The primary cause of serious complication is needle misplacement.

The eye:

- The risk of globe perforation and rupture is 1 in 350 to 7 in 50,000,[3] increasing with axial lengths greater than 26 mm.
- Arterial puncture may lead to retrobulbar haemorrhage and compressive haematoma, which may threaten retinal perfusion.
- Optic nerve damage is rare but may result in blindness.
- Chemosis from subconjunctival spread of local anaesthetic.

Systemic:

- Arterial injection, leading to seizures.[4]
- Injection under the dura of the optic nerve, leading to a total spinal.
- Oculocardiac reflex from traction on the globe (trigeminal afferent; vagal efferent), producing bradycardia.

Post-procedure checks

The signs of a successful block include:

- Ptosis.
- Akinesia.
- Inability to close the eye once opened.

References

1. Katz J, Feldman MA, Bass EB, *et al.* Adverse intraoperative medical events and their association with anesthesia management strategies in cataract surgery. *Ophthalmology* 2001; **108**: 1721–6.
2. Liu C, Youl B, Mosely I. *et al.* Magnetic resonance imaging of the optic nerve in the extremes of gaze. Implications for the positioning of the globe for retrobulbar anaesthesia. *Br J Ophthalmol* 1992; **76**: 728–33.
3. Edge R, Navon S. Scleral perforation during retrobulbar and peribulbar anesthesia: Risk factor and outcome in 50,000 consecutive injections. *J Cataract Refract Surg* 1999; **25**: 1237–44.
4. Aldrete J, Romo-Salas F, Arora S, *et al.* Reverse arterial blood flow as a pathway for central nervous system toxic responses following injection of local anaesthetics. *Anesth Analg* 1978; **57**: 428–33.

The nose

1. Overview

The part of the respiratory tract superior to the hard palate.

Functions:

- Olfaction.
- Breathing.
- Filtration and humidification of inhaled substances.
- Reception of secretions from the paranasal sinuses and nasolacrimal ducts.

Two nasal cavities are separated by the nasal septum. The cavities open anteriorly at the nares and posteriorly into the nasopharynx at the choanae.

2. The skeleton of the nose

The skeleton of the nose consists of a superior bony portion and an inferior cartilaginous portion (Figures 3.19 and 3.20).

Note that the superior and middle conchae are part of the ethmoid bone; the inferior concha is a bone in its own right.

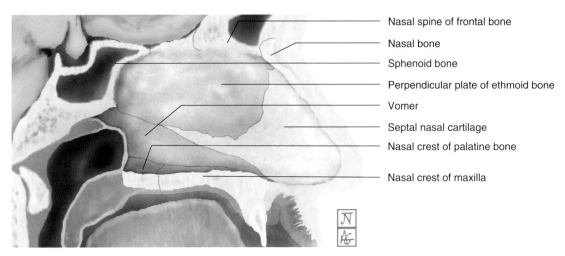

Nasal spine of frontal bone

Nasal bone

Sphenoid bone

Perpendicular plate of ethmoid bone

Vomer

Septal nasal cartilage

Nasal crest of palatine bone

Nasal crest of maxilla

Figure 3.19 Anatomy of the nasal septum. Note that the septum comprises seven bones and the nasal cartilage. The vomer is a bone in its own right.

a

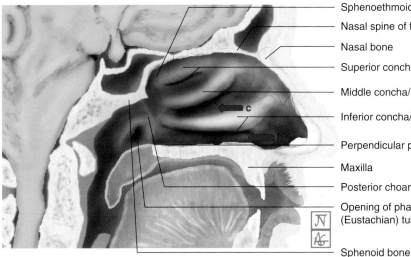

Sphenoethmoidal recess

Nasal spine of frontal bone

Nasal bone

Superior concha/ superior meatus

Middle concha/ middle meatus

Inferior concha/ inferior meatus

Perpendicular plate of palatine bone

Maxilla

Posterior choana

Opening of pharyngotympanic (Eustachian) tube

Sphenoid bone

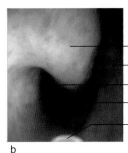

Inferior meatal approach (bottom red arrow)

Inferior concha

Septum

Inferior meatus

Floor of nose

Tip of epidural catheter for "spray as you go" technique

b

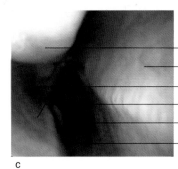

Middle meatal approach (top red arrow)

Middle concha

Septum (deformed)

Middle meatus

Inferior concha

Posterior choana

Floor of nose

c

Figure 3.20 (a) Anatomy of the lateral wall of the nose. Note that the lateral wall is formed of seven bones. The superior and middle conchae are parts of the ethmoid bone; the inferior concha is a bone in its own right. Endoscopic access to the airway is usually undertaken via the inferior meatus (lower arrow and view b), but a middle meatal approach (upper arrow and view c) may be undertaken if the inferior meatus is unfavourable.

3. The nasal cavities

The cavity is lined with mucosa, except at the vestibule, which is lined with skin and vibrissae.

The inferior two-thirds of the mucosa is the respiratory area; the superior one-third is the olfactory area (which contains olfactory cells).

Boundaries of the nasal cavities:

i) Roof

Curved and narrow.
Formed from the nasal, frontal, ethmoid and sphenoid bones.

ii) Floor

Palatine process of maxilla and horizontal plate of palatine bones.

iii) Lateral wall

Conchae (superior, middle and inferior).
Maxilla, palatine and ethmoid bones.

iv) Medial wall

Nasal septum, formed from the septal cartilage and the ethmoid, vomer, palatine and maxillary bones.

The nasal conchae divide the nasal cavity into four passages, into which drain the adjacent sinuses:

i) Sphenoethmoidal recess

Receives the sphenoidal sinus.

ii) Superior meatus

Receives the posterior ethmoidal sinus.

iii) Middle meatus

Receives the frontal, ethmoidal and maxillary sinuses.

iv) Inferior meatus

Receives the nasolacrimal duct (tears from the eye).

This is the largest passage and so is often the route of choice during a nasal fibreoptic intubation. The operator aims to guide the scope between the floor of the nasal cavity and the inferior concha (Figure 3.20).

4. Neurovascular supply

i) Arterial

- Anterior and posterior ethmoidal arteries (from the ophthalmic branch of the internal carotid artery).
- Sphenopalatine artery (from the maxillary branch of the external carotid).
- Greater palatine artery (from the maxillary branch of the external carotid).
- Superior labial (from the facial branch of the external carotid).
  All five arteries anastomose in Little's area at the anteroinferior part of the septum (from where most epistaxes originate).
  It can be seen from the origins of the above vessels that in severe epistaxis ligation of the external carotid artery may not arrest the flow of blood.

ii) Venous

A rich venous plexus drains to the internal jugular veins (via the sphenopalatine or facial veins) or to the cavernous sinus (via the ophthalmic veins).

iii) Nervous

Special sense: CN I in the olfactory area (superior one-third).
Sensation: (see Figure 3.21)
Anterosuperiorly – CN V_1 (anterior and posterior ethmoidal nerves).
Posteroinferiorly – CN V_2 (nasopalatine and greater palatine nerves).

Regional anaesthesia for awake fibreoptic intubation (AFOI)

Introduction

Fibreoptic laryngobronchoscopic intubation is a core technique for the management of a difficult airway. Effective anaesthesia of the airway is required to perform the technique safely in an awake patient.[1]

Indications

Previous or suspected difficult airway or difficult mask ventilation, particularly if the front-of-neck rescue strategy is likely to be difficult or impossible; previous AFOI; and to avoid iatrogenic injury in unstable cervical spine injury.

Specific contraindications

Absolute:
- Patient refusal.
- Operator not familiar with the technique.
- Allergy to local anaesthetic agents.

Relative:
- Contamination of the upper airway – blood, abscess, friable tumour.
- Uncooperative patient.
- Grossly distorted anatomy – as it may be difficult to pass the tube because of 'hold up' despite a successful bronchoscopy.
- Fractured base of skull in nasal approach.
- Penetrating eye injuries.

Pre-procedure checks

- Familiarity with technique.
- Monitoring as per AAGBI guidelines.
- IV access.
- Resuscitation drugs and equipment.
- Trained assistant.
- Titratable sedation may be required, preferably delegated to a second anaesthetist – e.g. remifentanil or propofol target-controlled infusion (TCI).
- Supplementary oxygen – 2 l/min to opposite nostril using nasal sponge.
- Consider the availability of an ENT or maxillofacial surgeon ready to perform a surgical airway in the event of procedural failure.

Technique

i) Topicalisation technique

A maximum dose of up to 9 mg/kg of lidocaine may be used,[2] based on lean body weight, made up of a combination of 4% lidocaine, 10% lidocaine and co-phenylcaine (5% lidocaine with 0.5% phenylephrine).
There are many recipes for topicalisation, and one example is given here.[3]
The patient is positioned comfortably in a semi-reclining or supine position.
- 4 ml of 4% lidocaine via nebuliser.
- 3 mcg/kg glycopyrolate is given IV to reduce secretions.
- A vasoconstrictor such as xylometazoline (Otrivine) or cocaine (up to 1.5 mg/kg, made up as 1 ml 10% cocaine, 1 ml of 1:1000 adrenaline and 8 ml 0.9% saline) is applied.
- 2.5 ml of co-phenylcaine is applied to the inferior nasal meatus (Figure 3.20b), turbinates and posterior nasal space with a cotton bud mounted on a stick or a mucosal atomiser device.
- 4 sprays of 10% lidocaine, applied to the oropharynx, aiming for the glottis, using the atomiser.

The remaining local anaesthetic dose is administered in 1 ml aliquots of 4% lidocaine via an epidural catheter passed through the working channel of the endoscope ('spray as you go' technique), specifically targeting the turbinates, and the area above (the vestibule) and below (infraglottic cavity) the vocal cords.

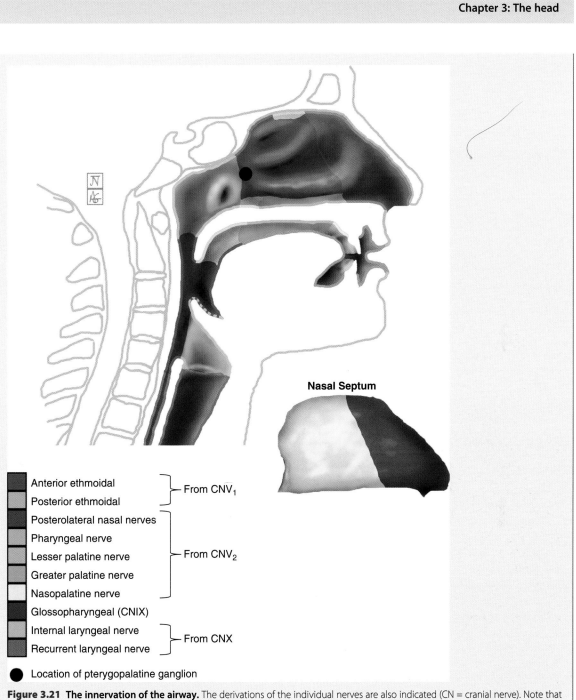

Nasal Septum

Anterior ethmoidal
Posterior ethmoidal
⎤ From CN$\overline{V}_1$

Posterolateral nasal nerves
Pharyngeal nerve
Lesser palatine nerve
Greater palatine nerve
Nasopalatine nerve
⎤ From CNV_2

Glossopharyngeal (CNIX)

Internal laryngeal nerve
Recurrent laryngeal nerve
⎤ From CNX

● Location of pterygopalatine ganglion

Figure 3.21 The innervation of the airway. The derivations of the individual nerves are also indicated (CN = cranial nerve). Note that the anterior surface of the epiglottis is innervated by the the glossopharyngeal nerve, the internal laryngeal nerve, or a combination of the two. The insert shows the right side of the nasal septum.

ii) Techniques for blocking individual nerves of the airway

Anterior and posterior ethmoidal nerve block
Blocked by a cotton bud mounted on a stick and soaked in local anaesthetic applied to the roof of the nose, by the anterior cribriform plate, and left for 5–10 minutes.

Greater palatine, lesser palatine, posterolateral nasal nerves, pharyngeal nerve and nasopalatine nerve block

Blocked at the pterygopalatine ganglion. A cotton bud mounted on a stick and soaked in local anaesthetic is passed along the upper border of the middle turbinate to the posterior wall of the nasopharynx, and left for 5–10 minutes.

Glossopharyngeal nerve block

Blocked intraorally by injecting 2–5 ml of local anaesthetic with a 22G needle submucosally at the base of the palatoglossal fold. It may also be blocked extraorally with a peristyloid approach, but the proximity of the carotid artery makes this approach undesirable.

Superior laryngeal nerve block

Blocked by bilateral local anaesthetic injection adjacent to the greater cornu of the hyoid bone. The patient is supine with head extended, and the cornu of the hyoid is identified by palpating outward from the thyroid notch along the upper border of the thyroid cartilage until the cornu is felt just superior to its posterolateral margin. A 25G needle is used to contact the lateral aspect of the greater cornu and withdrawn slightly before aspirating and then injecting 2 ml 2% lidocaine. An alternative is to place pledgets soaked in local anaesthetic bilaterally in the piriform fossa with Jackson–Krause forceps after topicalisation of the pharynx.

Recurrent laryngeal nerve block

Blocked by transtracheal injection. A 20–22G needle is inserted perpendicularly through the cricothyroid membrane into the trachea until air is aspirated, and 2 ml 2% lidocaine is injected during exhalation. The resultant cough spreads local anaesthetic within the tracheo-bronchial tree. A larger-gauge needle may facilitate more rapid delivery of local anaesthetic and gives the user the option of passing the wire from a Melker cricothyroidotomy set (see box on *Surgical and cannula cricothyroidotomy*).

Complications[2]

- Failure, which may result in loss of the airway and death if a suitable alternative plan is not in place.
- Symptoms of systemic local anaesthesia (dizziness, dysphoria, circumoral tingling etc.).

Post-procedure checks

Before inducing general anaesthesia:
- Confirm correct placement of the endotracheal tube with capnography, auscultation of bilateral air entry and observation of bilateral chest movement.

Prior to extubation follow the Difficult Airway Society (DAS) Extubation Guidelines 'at risk' algorithm.[4]

References

1. Cook T, Woodall N, Ferk C, *et al*. Fourth National Audit Project. Major complications of airway management in the UK: results of the Fourth National Audit Project of the Royal College of Anaesthetists and the Difficult Airway Society. *Anaesthesia* 2011; **106**: 617–42.
2. Woodall N, Harwood R, Barker G. Complications of awake fibreoptic intubation without sedation in 200 healthy anaesthetists attending a training course. *Br J Anaesth* 2008; **100**: 850–5.
3. Williams KA, Barker GL, Harwood RJ, Woodall NM. Combined nebulization and spray-as-you-go topical local anaesthesia of the airway. *Br J Anaesth* 2005; **95**: 549–53.
4. Popat M, Mitchell V, Dravid R, *et al*. Difficult Airway Society guidelines for the management of tracheal extubation. *Anaesthesia* 2012; **67**: 318–40.

The mouth

1. Overview

The mouth extends from the lips anteriorly, to the oropharyngeal isthmus posteriorly, where it opens into the oropharynx.
It is subdivided into:

i) The vestibule

Slit-like cavity between the teeth/gingivae and the lips/cheeks.

ii) The oral cavity

The space between the dental arches.
It is limited by:
- The palate (superiorly).
- The tongue (inferiorly).
- The oropharyngeal isthmus formed by the palatoglossal and palatopharyngeal arches (known as the pillars of fauces) posteriorly, where it communicates with the oropharynx.

When the mouth is closed, this space is fully occupied by the tongue.

2. The palate

i) Hard palate

Formed by the maxilla (anteriorly) and the palatine bone (posteriorly).

ii) Soft palate

Hangs like a curtain from the hard palate, with the uvula in the midline.
The skeleton is formed by a fibrous aponeurosis.
During swallowing the soft palate initially facilitates movement of the food bolus towards the back of the mouth. It then swings posterosuperiorly to prevent food entering the nasopharynx.

3. The tongue

Divided into anterior and posterior parts by the V-shaped terminal sulcus. The main features and innervation are shown in Figure 3.22.
The anterior part lies in the oral cavity; the posterior part lies in the oropharynx.
Inferiorly is the lingual frenulum, either side of which are the openings of the submandibular ducts and the lingual veins.

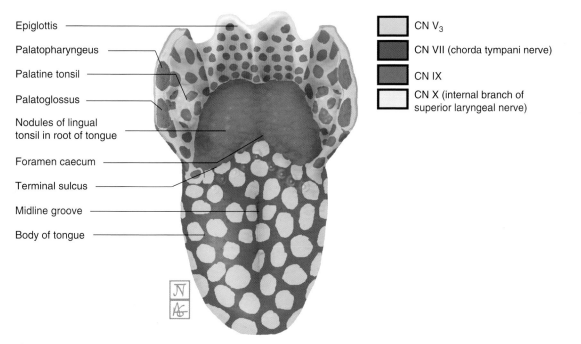

Epiglottis
Palatopharyngeus
Palatine tonsil
Palatoglossus
Nodules of lingual tonsil in root of tongue
Foramen caecum
Terminal sulcus
Midline groove
Body of tongue

CN V$_3$
CN VII (chorda tympani nerve)
CN IX
CN X (internal branch of superior laryngeal nerve)

Figure 3.22 Anatomical features and innervation of the tongue. CN, cranial nerve. Note that the chorda tympani nerve carries sensation of taste (special sensory), whereas CN V$_3$ carries general sensation (e.g. touch, temperature, etc.). CN IX and X carry both taste and general sensation. The anterior portion of the epiglottis is innervated by CN IX, CN X or a combination of the two.

4. Neurovascular supply

All vessels are derived from the external carotid arteries and drain into the internal jugular veins.

The sublingual veins (tributaries of the lingual veins), found on either side of the frenulum, offer a route for drug administration which avoids first-pass metabolism.

i) Arterial and venous

- Vestibule: facial arteries (labial branches) and veins.
- Teeth: maxillary arteries (alveolar branches) and veins.
- Tongue: lingual arteries (direct from the external carotid) and veins.
- Palate: maxillary arteries (greater and lesser palatine branches) and veins.

ii) Nervous

- Vestibule:
  Motor – CN VII.
  Sensory – CN V_2 and CN V_3.
- Teeth:
  Sensory – maxillary teeth from CN V_2 and mandibular teeth from CN V_3.
- Tongue:
  Motor – CN XII (except palatoglossus, supplied by CN X).
  Special sensory – Taste to the anterior two-thirds from the chorda tympani nerve (branch of CN VII); to the posterior one-third from CN IX.
  Sensory – to the anterior two-thirds from CN V_3; to the posterior one-third from CN IX.
- Palate:
  Motor – CN X.
  Sensory – CN V_2.

The neck

Fascia of the neck

Structures within the neck are compartmentalised by fascial layers, which determine the spread of infection and limit the diffusion of deposited local anaesthetic agents (Figure 4.1).

1. Superfical cervical fascia

A subcutaneous layer which envelops platysma muscle anteriorly.

2. Deep cervical fascia

Consists of three fascial layers:

i) Investing

Invests trapezius posteriorly and sternocleidomastoid anteriorly.
Attaches superiorly to the skull, the zygomatic arches, the mandible, the hyoid bone and spinous processes of the cervical vertebrae.
Attaches inferiorly to the clavicles and acromions/spines of scapulae.

ii) Pretracheal

Encloses the trachea, oesophagus, thyroid gland and infrahyoid muscles.
Blends superiorly with the buccopharyngeal fascia of the pharynx and inferiorly with the fibrous pericardium.

This communication explains why some pharyngeal abscesses may track into the mediastinum. They warrant not only CT head/neck/chest but also consideration of panendoscopy to identify the infectious source. The largest interfascial space often implicated in such infections is the retropharyngeal space – the potential space between the alar fascia and the prevertebral fascia (Figure 4.1).

iii) Prevertebral

Encloses the vertebral column and its associated muscles.
Extends from the base of the skull to T3, where it merges with the anterior longitudinal ligament.
Forms:

- The carotid sheath, which contains arteries (common and internal carotid), a vein (internal jugular), nerves (vagus, sympathetic fibres, carotid sinus nerve) and lymph nodes. Also communicates with the mediastinum, forming a potential route for extravasated blood from vascular puncture.
- The sheath of the brachial plexus, which begins from the tubercles of the cervical vertebrae and extends as far laterally as the axilla. Offers the potential for easy spread of local anaesthetic deposited within it.

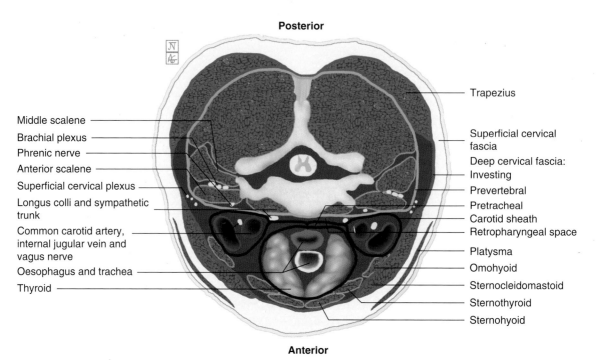

Figure 4.1 Transverse section of the neck at the level of C6. The alar fascia unites the two carotid sheaths (unlabelled); the space between the alar fascia and the prevertebral fascia is the retropharyngeal space.

The vessels of the neck

1. Arteries

The common carotid arteries arise from the brachio-cephalic trunk behind the sternoclavicular joint (on the right) and directly from the aorta (on the left). They ascend in the neck in the carotid sheath, dividing into internal and external carotid arteries opposite the upper border of the thyroid cartilage (C4), the site of the carotid body (a chemoreceptor).

i) Internal carotid artery

No branches in the neck.
Proximally, houses the carotid sinus (a baroreceptor). Ascends medial to the internal jugular vein to pass through the carotid canal of the temporal bone and across the foramen lacerum.
Terminates by dividing into the anterior and middle cerebral arteries (Figure 3.6).

ii) External carotid artery

Supplies structures of the neck and the head external to the skull (with the exception of the middle meningeal artery).

Has eight branches (from inferior to superior): ascending pharyngeal, superior thyroid, lingual, facial, occipital, posterior auricular, maxillary and superficial temporal (palpable anterior to the tragus of the ear). Mnemonic – Always Sleep Lying Flat Or People May Stare.

2. Veins

i) Internal jugular vein

Drains the brain, anterior face, viscera and deep muscles of the neck.
The continuation of the sigmoid sinus at the jugular foramen.
Descends in the carotid sheath lateral to the internal then the common carotid arteries.
Passes deep to the space between the sternal and clavicular heads of sternocleidomastoid, where it is accessible for cannulation ('low approach').
Unites with the subclavian vein posterior to the sternal end of the clavicle to form the brachiocephalic vein (Figure 5.25).
Has superior and inferior bulbs. The inferior bulb contains a bicuspid valve, preventing retrograde

flow of venous blood, but also potentially obstructing passage of a central line wire.

ii) External jugular vein

Drains the scalp and lateral face.
Runs from the angle of the mandible, across sternocleidomastoid.

Pierces the investing layer of deep fascia and drains into the subclavian vein superior to the clavicle.
A useful site for vascular access in emergencies such as cardiac arrest, where poor cardiac function results in venous distension.

Internal jugular venous access

Introduction

The internal jugular (IJ) vein is located lateral to the common and internal carotid arteries and deep to the space between the sternal and clavicular heads of the sternocleidomastoid muscle (Figure 5.25).

Indications

Vascular access, central venous pressure monitoring, cardiac output measurement, drug administration, renal replacement therapy and plasmapheresis. The internal jugular vein may offer the best balance of risk of pneumothorax, haemothorax, malposition, incidence of infection and arterial puncture when compared to the subclavian or femoral approaches.[1]

Specific contraindications

Absolute:
- Patient refusal.
- Cutaneous infection at insertion site.

Relative:
- Coagulopathy.
- Uncooperative patient.
- Respiratory decompensation – where lying flat would precipitate a catastrophic decline in the patient's state.

Pre-procedure checks

- Airway and ventilation equipment.
- IV access (if possible).
- Resuscitation drugs and equipment.
- Trained assistant.
- Sedation and analgesia may be required.
- Supplementary oxygen.
- Full aseptic technique (gown/mask/gloves/hat). Meticulous skin prep with 2% chlorhexidine gluconate in 70% isopropyl alcohol, allowed to dry fully.[1,2]

Technique

Supine, 15–30° head down Trendelenberg, with head turned 45° away from insertion site.

Aseptic Seldinger technique
The central venous catheter is flushed with 0.9% saline and the lumens clamped except for the distal port, through which will pass the guidewire.
A subcutaneous wheal of local anaesthetic is injected at the insertion site with a 20G needle. The dedicated Seldinger needle is advanced towards the central vein, preferably under ultrasound guidance,[3] at 30–40° to the skin while continuously aspirating, until venous blood flows freely into the syringe. The needle is flattened out to allow easier passage of the guidewire. The syringe is disconnected from the needle and the presence of venous blood confirmed. The needle is held steady and 15 cm of guidewire is inserted while watching the ECG. If a dysrhythmia is provoked the wire is withdrawn slightly. The needle is removed with the guidewire held in place. A knife is used to nick the skin alongside the guidewire. The dilator is threaded over the wire and the skin and tract dilated.
Sterile gauze is placed over the puncture site as the dilator is removed. The catheter is then passed over the guidewire and the guidewire removed. The distal port is clamped swiftly to avoid air embolus. A sterile syringe containing 0.9%

saline is used to aspirate and flush each port, confirming venous placement. The catheter is sutured in place in accordance with the manufacturer's instructions and a sterile, clear dressing is placed over the insertion site.

Landmark

The insertion point is halfway along a line drawn from the mastoid process to the sternal notch (high approach) or at the apex of the triangle formed by the two heads of the sternocleidomastoid muscle (low approach).
The needle is inserted just lateral to the carotid pulse at 30–40° to the skin and directed towards the ipsilateral nipple. The vein should be located within a few centimetres of the skin, and caution is advised against inserting the needle deeper than this, especially with the low approach.

Ultrasound-assisted block

A high-frequency probe is placed in a transverse plane halfway along a line drawn from the mastoid process to the sternal notch. The internal jugular vein is visualised lateral to the carotid artery. The vein is usually larger, oval-shaped and compressible; the artery is pulsatile, usually smaller, more round and not easily compressible. Doppler will further help to differentiate the vessels. Difficulty in visualising the vein may be due to hypovolaemia or excessive probe pressure.
The vein may be approached either out-of-plane or in-plane. The out-of-plane approach allows visualisation of both vessels simultaneously, although needle appreciation, even with a relatively vertical needle intent, is limited. The in-plane approach allows good needle visualisation, but only one vessel is seen at any one time (and so this approach is recommended only for experienced practitioners). Following vessel puncture and aspiration of blood, the needle is flattened out (as for landmark approach) to allow easier passage of the guidewire. Once the guidewire has been passed an in-plane ultrasound assessment is made to confirm correct positioning within the vein prior to dilation.

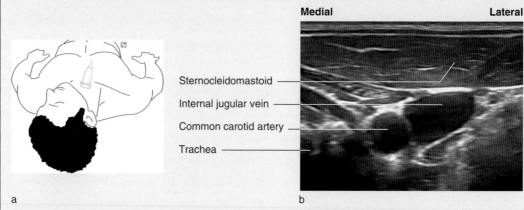

Medial   **Lateral**

Sternocleidomastoid

Internal jugular vein

Common carotid artery

Trachea

a   b

Figure 4.2 Ultrasound appearance of the out-of-plane approach to the internal jugular vein.

Complications

- Pneumothorax (0.1–0.2%). The internal jugular approach has a lower incidence of pneumothorax than the subclavian approach, but has a higher rate of infection because of the proximity of respiratory and oropharyngeal secretions.
- Catheter-related bloodstream infection.
- Arterial puncture and haemorrhage.
- Thrombosis.
- Arrhythmias during wire insertion.
- Air embolus.
- Loss of the guidewire into the cavity.
- Thoracic duct injury on left side.
- Nerve injury.

Post-procedure checks

- Chest x-ray to confirm line position (target: beneath the clavicle and outside the cardiac shadow) and exclude pneumothorax.
- Continual reassessment for catheter requirement, and routine hand washing.[1]

References

1. Pronovost P, Needham D, Berenholtz S, *et al*. An intervention to decrease catheter-related bloodstream infections in the ICU. *N Engl J Med* 2006; **355**: 2725–32.
2. Pratt RJ, Pellowe CM, Wilson JA *et al*. Epic2: National evidence-based guidelines for preventing healthcare-associated infections in NHS hospitals in England. *J Hosp Infect* 2007; **65**: S1–64.
3. National Institute for Clinical Excellence. *Guidance on the use of ultrasound locating devices for placing central venous catheters*. Technology Appraisal 49. London: NICE; 2002. http: //guidance.nice.org.uk/TA49 (accessed November 2013).

The nerves of the neck

1. The cervical plexus

Formed from the ventral rami of C1–4.
Supplies:

- Motor – the muscles of the neck and diaphragm.
- Sensory – the skin of the head and neck (except C1, which is an entirely motor nerve).

The ventral rami of C1–4 pass along their respective transverse processes, where they receive grey rami communicantes from the superior cervical ganglion (see Chapter 1, section 6).

They pass behind the vertebral artery before reaching the tips of the transverse processes, where they divide into ascending and descending branches.

The branches from adjacent rami unite, forming nerve loops.

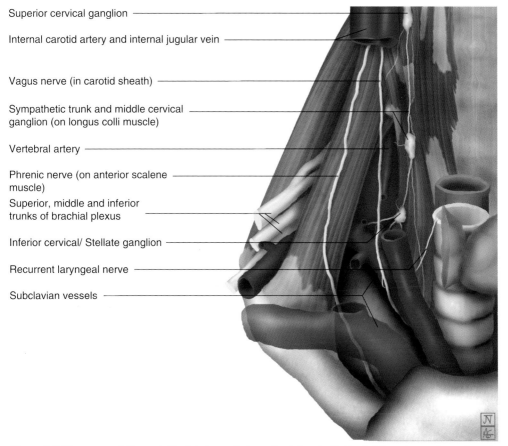

Superior cervical ganglion

Internal carotid artery and internal jugular vein

Vagus nerve (in carotid sheath)

Sympathetic trunk and middle cervical ganglion (on longus colli muscle)

Vertebral artery

Phrenic nerve (on anterior scalene muscle)

Superior, middle and inferior trunks of brachial plexus

Inferior cervical/ Stellate ganglion

Recurrent laryngeal nerve

Subclavian vessels

Figure 4.3 The nerves of the neck. The inferior cervical ganglion is fused with the first thoracic ganglion to form the stellate ganglion in 80% of patients. The cervical plexus has been omitted for clarity.

The loops and the nerves leaving them form the cervical plexus, which is found anterior to middle scalene and posterior to sternocleidomastoid.

Four major branches emerge from the plexus (Figure 4.4):

i) Deep cervical plexus

Entirely motor neurons, which emerge anteriorly from the plexus.

Supplies the anterior vertebral muscles with small additional contributions to the large neck muscles (sternocleidomastoid, trapezius, middle scalene and levator scapulae). Also emerging from the deep plexus is the ansa cervicalis, which supplies the infrahyoid muscles.

ii) Superficial cervical plexus

Entirely sensory neurons, which emerge posteriorly from the plexus at the nerve root of the neck (the middle of sternocleidomastoid at its posterior border; Erb's point).

The nerves, which supply the cutaneous innervation to the neck and shoulder, are (Figure 4.5):

- Lesser occipital (C2).

- Great auricular (C2, 3).
- Transverse cervical (C2, 3).
- Supraclavicular (C3, 4). Divides into medial, intermediate and lateral branches. Importantly, the lateral branch provides cutaneous innervation to the anterior shoulder.

iii) Phrenic nerve

Supplies motor and sensory fibres to the diaphragm.

Derived from ventral rami of C3–5 ('C3, 4, 5 keeps the diaphragm alive'), which unite at the lateral edge of anterior scalene muscle.

Descends on anterior scalene and enters the thorax posterior to the internal jugular vein.

The proximity of the phrenic nerve to the brachial plexus makes it prone to anaesthesia during interscalene block (incidence ≈ 85%). This has particular implications for those with limited respiratory function.

iv) Communicating branches

Pass to the hypoglossal and vagus nerves and the sympathetic trunks.

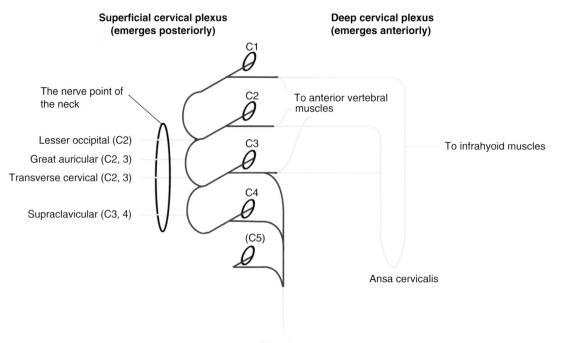

Superficial cervical plexus
(emerges posteriorly)

Deep cervical plexus
(emerges anteriorly)

C1

The nerve point of the neck

C2 — To anterior vertebral muscles

Lesser occipital (C2)

Great auricular (C2, 3)

C3 — To infrahyoid muscles

Transverse cervical (C2, 3)

C4

Supraclavicular (C3, 4)

(C5)

Ansa cervicalis

Phrenic nerve

Figure 4.4 Schematic illustration of the cervical plexus. C5 is not part of the cervical plexus and so is shown in brackets. The nerve point of the neck is the middle of sternocleidomastoid at its posterior border.

Superficial, intermediate and deep cervical plexus block

Introduction

The cutaneous innervation of the neck, supraclavicular area and shoulder is supplied by the superficial cervical plexus. The deep cervical plexus supplies motor innervation to the neck (see text).

Indications

Cutaneous analgesia for procedures involving the anterior neck and anterior shoulder.

There is a trend towards using superficial or intermediate block in preference to deep cervical plexus block for awake carotid endarterectomy, as there is no difference in requirement for perioperative local anaesthetic or patient satisfaction. Additional local anaesthetic may be injected subcutaneously along the incision site to block contralateral innervation, around the angle of the jaw to tolerate a skin retractor, and within the carotid artery sheath intra-operatively to block the glossopharyngeal nerve.

Technique

Landmark

Superficial cervical plexus block

The patient is in a supine position with the head turned away from the side to be blocked. A 22G needle is inserted to a maximum depth of 5 mm at the midpoint of the posterior border of sternocleidomastoid (at approximately C6 – the level of the cricoid cartilage) and advanced superiorly and inferiorly. After negative aspiration, 10–20 ml of subcutaneous local anaesthetic is injected along the entire length of the muscle in this plane to form a 'sausage' encompassing the posterior and anterior borders of sternocleidomastoid.

There may be contralateral innervation from the opposite side of the neck, so a further 5 ml of local anaesthetic is injected subcutaneously along the midline from the thyroid cartilage to the suprasternal notch (prevents discomfort from skin retractors).

Intermediate cervical plexus block

A 22G needle is inserted perpendicularly to the skin at the midpoint of the posterior border of sternocleidomastoid until a loss of resistance or 'pop' is felt as the investing fascia is penetrated at a depth of 1–2 cm. The needle is fixed in this position and 10–20 ml of local anaesthetic is injected in aliquots after negative aspiration. The investing fascia of the neck may be incomplete, resulting in a similar distribution of local anaesthetic to a superficial injection.[1,2]

Deep cervical plexus block

The patient is in a supine position with the head turned away from the side to be blocked. The posterior border of sternocleidomastoid is palpated at the level of the thyroid cartilage (C4) and the fingers are moved laterally into the interscalene groove. The skin is stabilised between two fingers and a 25–50 mm block needle is inserted perpendicularly to the skin and advanced towards the contralateral elbow. The needle is advanced slowly, to a maximum depth of 2.5 cm, until the transverse process is contacted or paraesthesia elicited. The needle is withdrawn a few millimetres and 8–10 ml of local anaesthetic is injected in aliquots after negative aspiration.

Ultrasound-assisted block

Intermediate cervical plexus block

The patient is in a lateral position with the side to be blocked uppermost. A high-frequency probe is placed transversely over the posterior border of the midpoint of sternocleidomastoid, with the tapering posterior edge of the muscle in the middle of the screen. The brachial plexus and interscalene muscles are visualised. The plexus may be visualised as small hypoechoic structures arranged in a medial-to-lateral plane deep to the aponeurosis of the sternocleidomastoid muscle and superficial to the interscalene groove. The needle is inserted in-plane or out-of-plane and the tip placed adjacent to the plexus.

Hydrolocation (aspiration followed by injection of a few millilitres of local anaesthetic) may be used to confirm the correct position and then 10–15 ml of local anaesthetic injected in aliquots after negative aspiration. If the plexus cannot be visualised, deposition of local anaesthetic in this plane deep to sternocleidomastoid will provide reliable anaesthesia.

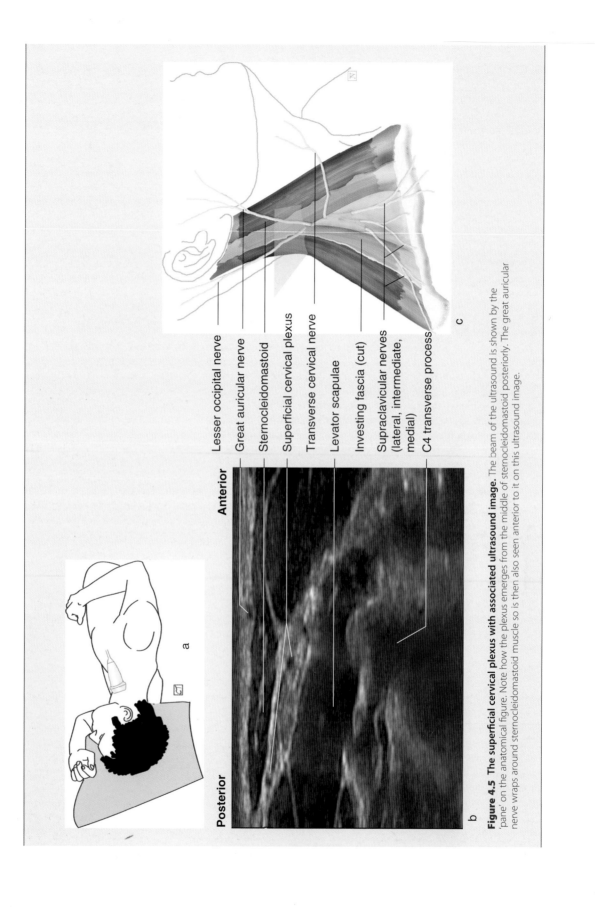

Posterior

Anterior

Lesser occipital nerve

Great auricular nerve

Sternocleidomastoid

Superficial cervical plexus

Transverse cervical nerve

Levator scapulae

Investing fascia (cut)

Supraclavicular nerves
(lateral, intermediate,
medial)

C4 transverse process

a

b

c

Figure 4.5 The superficial cervical plexus with associated ultrasound image. The beam of the ultrasound is shown by the 'pane' on the anatomical figure. Note how the plexus emerges from the middle of sternocleidomastoid posteriorly. The great auricular nerve wraps around sternocleidomastoid muscle so is then also seen anterior to it on this ultrasound image.

Complications

A systematic review could find no description of a serious complication related to superficial or intermediate cervical block in 30 years of publications,[3] whereas deep cervical block was associated with subarachnoid injection, intravascular injection and respiratory distress due to phrenic or laryngeal nerve block.

- Phrenic nerve block.
- Recurrent laryngeal nerve block.
- Stellate ganglion block.
- Haematoma.
- Local anaesthetic toxicity due to inadvertent intravascular injection.
- Spinal anaesthesia – accidental injection within the dural sleeve that accompanies the cervical nerve roots.

References

1. Nash L, Nicholson H, Zhang M. Does the investing layer of the deep cervical fascia exist? *Anesthesiology* 2005; **103**: 962–8.
2. Ramachandran S, Picton P, Shanks A, *et al.* Comparison of intermediate vs. subcutaneous cervical plexus block for carotid endarterectomy. *Br J Anaesth* 2011; **107**: 157–63.
3. Pandit J, Satya-Krishna R, Gration P. Superficial or deep cervical plexus block for carotid endarterectomy: a systematic review of complications. *Br J Anaesth* 2007; **99**: 159–69.

2. The brachial plexus

Formed from the ventral rami of C5–8 and T1 (occasionally C4–8 or C6–T2).
Supplies:

- Motor and sensory innervation to the upper limb and shoulder joint.

The plexus has the following components (Figure 4.6):

i) Roots

Emerge from the intervertebral foramina, then pass between anterior and middle scalene muscles.

ii) Trunks

- Superior – union of C5 and C6 roots.
- Middle – continuation of C7 root.
- Inferior – union of C8 and T1 roots.
  Emerge between the anterior and middle scalene muscles and pass across the posterior triangle of the neck (where they may be palpated in thin subjects) towards the first rib.

iii) Divisions

Form at the lateral border of the first rib.
Each trunk divides into an anterior and posterior division behind the clavicle. The anterior divisions supply the anterior (flexor) compartments of the upper limb, the posterior divisions supply the posterior (extensor) compartments.

The divisions continue towards the axilla.

iv) Cords

Form in the axilla and named according to their position in relation to the axillary artery:

- Lateral – union of the anterior divisions of the superior and middle trunks.
- Medial – continuation of the anterior division of the inferior trunk.
- Posterior – union of the posterior divisions of all three trunks.

v) Terminal branches

The terminal nerves of the arm (distal to the shoulder) and forearm are described in detail in Chapter 6. They emerge from the plexus as follows:

- Musculocutaneous nerve – from the lateral cord.
- Ulnar nerve – from the medial cord.
- Median nerve – from the lateral and medial cords.
- Radial and axillary nerves – from the posterior cord.
- Axillary nerve – from the posterior cord.

The suprascapular nerve is derived from the superior trunk. It is the main sensory nerve to the posterior and superior aspect of the shoulder joint, as well as the acromioclavicular joint, associated bursa and ligaments, and the supra- and infraspinatous muscles. It does not

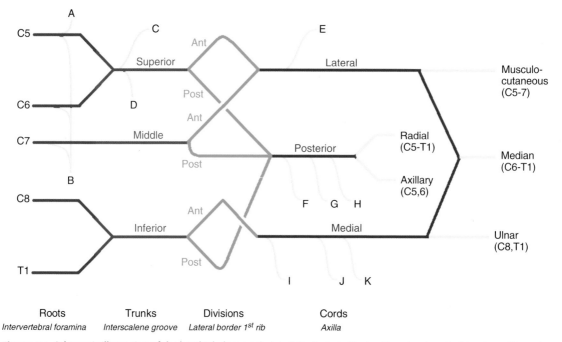

Roots   Trunks   Divisions   Cords

Intervertebral foramina   *Interscalene groove*   *Lateral border 1st rib*   *Axilla*

Figure 4.6 Schematic illustration of the brachial plexus. Labels in italics indicate the location of each part of the plexus. The major nerves are shown on the illustration and are discussed in the text. The lettered nerves (and their actions) are:
A – Dorsal scapular nerve (rhomboids, levator scapulae)
B – Long thoracic nerve (serratus anterior)
C – Suprascapular nerve (supra & infraspinatus and shoulder joint)
D – Nerve to subclavius (subclavius and sternoclavicular joint)
E – Lateral pectoral nerve (pectoralis major & minor)
F and H – Upper and lower subscapular nerves (subscapularis and teres major)
G – Thoracodorsal nerve (latissimus dorsi)
I – Medial pectoral nerve (pectoralis major & minor)
J – Medial cutaneous nerve of the arm (Figure 6.4)
K – Medial cutaneous nerve of the forearm (Figure 6.4)

usually provide *cutaneous* innervation to the scapula or posterior shoulder (this comes from the supraclavicular nerve from the cervical plexus – see Figure 6.5). Brachial plexus blockade for shoulder surgery must therefore be placed no more distally than the trunks (i.e. an interscalene block is required). For awake surgery, additional anaesthesia of the suprascapular nerve is required (which may be performed in isolation of the brachial plexus – see box on *Suprascapular nerve block*, below). The brachial plexus is surrounded by fatty tissue which is enclosed in a fibrous sheath extending from the transverse processes of the vertebrae to the axilla. Spread of local anaesthetic solutions is facilitated by deposition within the sheath.

Interscalene brachial plexus block

Introduction

Interscalene brachial plexus block is performed between the anterior and middle scalene muscles and produces anaesthesia of the shoulder, arm and elbow (Figure 4.7).
The shoulder joint is innervated by the axillary nerve and suprascapular nerve, and variably by contributions from the subscapular and musculocutaneous nerves. The *cutaneous* innervation of the shoulder is provided by the supraclavicular nerve (C3–4), a branch of the superficial cervical plexus (Figure 4.5c). Block of the supraclavicular nerve requires spread of local anaesthetic from the interscalene space or

a supplementary superficial cervical plexus block (see box on *Superficial, intermediate and deep cervical plexus block*).

Indications

Surgery on the shoulder, the lateral two-thirds of the clavicle and the proximal humerus. For awake arthroscopic procedures of the shoulder, anaesthesia of the anterior port site incisions requires supraclavicular nerve anaesthesia as described above.

Specific contraindications

Relative:
- Pre-existing neurological deficit in operative arm.
- Contralateral phrenic or recurrent laryngeal nerve paresis.
- Severe chronic obstructive airway disease (patients whose breathing requires use of accessory respiratory muscles) or severe asthma.

Technique

The patient is awake. Conscious sedation may be used.
8–30 ml of local anaesthetic is injected in aliquots after negative aspiration.

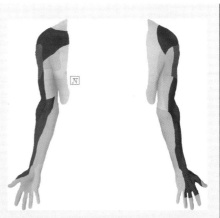

Figure 4.7 The distribution of anaesthesia following interscalene block. The area of anaesthesia is shown by the coloured areas, which correlate with Figure 6.5.

Landmark/nerve stimulator

The patient is in a supine position with the head turned away from the side to be blocked, to tense the sternocleido-mastoid muscle. With the head elevated, the index and middle fingers are placed immediately behind the lateral edge of sternocleidomastoid, over the belly of the anterior scalene muscle, and rolled laterally into the interscalene groove (Figure 4.8d). The patient may be asked to sniff to contract the scalene muscles and assist in palpation of the groove.

Winnie's approach[1]
A 50 mm block needle is inserted into the interscalene groove at the level of the cricoid cartilage (C6), aiming towards the contralateral elbow.

Meier's approach[2]
Useful when identification of the interscalene groove is difficult.
The needle is inserted 2 cm cranial to the cricoid cartilage at 30° to the skin, along a line drawn from the posterior border of sternocleidomastoid at the level of the thyroid prominence to the pulsation of the subclavian artery as it is palpated over the first rib.

Borgeat's approach[3]
This approach facilitates the placement of a catheter.
The needle is inserted 0.5 cm below the level of the cricoid cartilage at 45–60° to the skin and advanced caudally into and along the interscalene groove.[3]

Low interscalene approach
This is performed where the trunks lie close together, producing a more reliable block of the lower trunk and reducing the risk of cervical cord or vertebral artery puncture.
A 50 mm nerve stimulator needle is inserted 3–4 cm above the clavicle into the interscalene groove, perpendicular to the skin and with a slight caudal intent. The brachial plexus is very superficial, typically lying 1–2 cm below the skin, and for this reason the needle is not inserted deeper than 2.5 cm.

Nerve stimulation responses

- Deltoid – axillary nerve C5/6.
- Pectoralis – lateral pectoral nerve C5-7.
- Long head of triceps – axillary nerve C5/6.[4]

- Biceps – musculocutaneous nerve C5–7.
- Forearm or hand muscles.
- Phrenic nerve (which lies on the anterior scalene) – needle tip too anterior.
- Dorsal scapular nerve (within middle scalene) – needle tip too posterior.

Ultrasound-assisted block

The patient is supine for an out-of-plane approach, or lateral for an in-plane approach (Figure 4.8a,b). A high-frequency probe is placed in a transverse plane just below the level of the cricoid cartilage (C6) over the medial edge of the sternocleidomastoid. The carotid artery and thyroid are identified, and the probe is moved laterally towards the posterior border of sternocleidomastoid. The anterior and medial scalene muscles are identified deep to the tapering posterior edge of sternocleidomastoid, and the nerves of the brachial plexus are seen in the plane between them. Identification of the plexus may be aided by moving the probe to the supraclavicular fossa and tracking the plexus from its position anterolateral to the subclavian artery (see box on *Supraclavicular brachial plexus block*) back up the neck to the desired level.

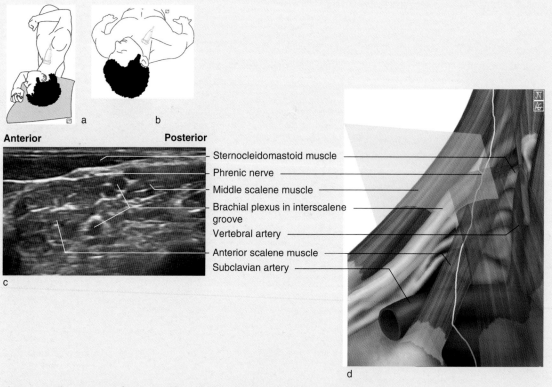

Anterior   **Posterior**

- Sternocleidomastoid muscle
- Phrenic nerve
- Middle scalene muscle
- Brachial plexus in interscalene groove
- Vertebral artery
- Anterior scalene muscle
- Subclavian artery

Figure 4.8 The anatomy of an interscalene block, with associated ultrasound image. The block may be performed in or out of plane, as shown by (a) and (b) respectively. See also Figure 4.9.

Local anaesthetic may be deposited in two locations (a combination of both may be used):
- Between the scalene muscles and the brachial plexus sheath. This requires 10–15 ml, which increases the likelihood of spread to the phrenic nerve (Figure 4.8d) but has a lower risk of damage to the plexus.
- Between the nerve roots within the brachial plexus sheath. There is a greater risk of nerve damage and sympathetic block (Horner's syndrome), but the smaller-volume injection (5 ml) limits spread to the phrenic nerve.

The needle approach for the out-of-plane block is perpendicular to the skin. The needle approach for the in-plane block is posterior to anterior through the middle scalene (care should be taken to avoid damage to the dorsal scapular nerve, which lies within the muscle). A 'pop' is often felt as the needle passes through the fascia on the medial aspect of the middle scalene. The injection is performed where the C5–7 nerves can be identified clearly and the needle trajectory avoids the vertebral artery and dural cuffs.

Ultrasound-assisted assessment of root level

The brachial plexus nerve root level can be identified with ultrasound. The hyperechoic surface reflection of the bony anterior and posterior tubercles of C4–6 can be reliably distinguished, with a hypoechoic nerve root held between them (Figure 4.9b). C7 is unlike the other cervical vertebrae as it lacks an anterior tubercle[5] and the vertebral artery does not pass through the foramen of the transverse process. It is identified by a larger posterior tubercle ending in a smooth down-sloping contour anteriorly, with the hypoechoic C7 root bordered only by the posterior tubercle. The vertebral artery can be identified with colour Doppler medial to the nerve root (Figure 4.9c).

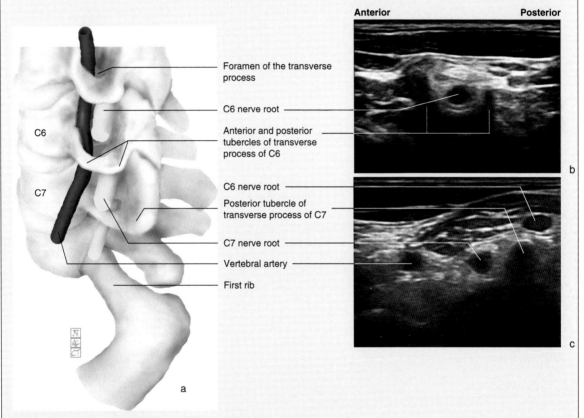

Figure 4.9 Anatomical and ultrasound images showing how the root level of the nerves of the brachial plexus are identified sonographically.

Complications

- Total spinal or epidural anaesthesia.
- Vertebral artery injection.
- Pneumothorax.
- Haematoma from carotid artery puncture.
- Local anaesthetic toxicity (inadvertent intravascular injection).
- Nerve injury (approximately 1–2 per 10,000).
- Horner's syndrome – less common with the low approach.
- Diaphragmatic paralysis – invariably present to some degree, depending on the volume of local anaesthetic used.

Post-procedure checks

- Inability to actively abduct the upper arm indicates blockade of the axillary nerve (C5/C6) and correlates with block success.[6]
- Assessment of extent of phrenic nerve block.

References

1. Winnie A. Interscalene brachial plexus block. *Anesth Analg* 1970; **49**: 455–66.
2. Meier G, Bauereis C, Heinrich C. Interscalene brachial plexus catheter for anaesthesia and postoperative pain therapy. Experience with a modified technique. *Anaesthetist* 1997; **46**: 715–19.
3. Borgeat A, Ekatodramis G. Anaesthesia for shoulder surgery. *Best Pract Res Clin Anaesthesiol* 2002; **16**: 211–25.
4. de Sèze MP, Rezzouk J, de Sèze M, *et al.* Does the motor branch of the long head of the triceps brachii arise from the radial nerve? *Surg Radiol Anat* 2004; **26**: 459–61.
5. Martinoli C, Bianchi S, Santacroce E, *et al.* Brachial plexus sonography: a technique for assessing the root level. *American Journal of Roentgenology* 2002; **179**: 699–702.
6. Wiener D, Speer K. The deltoid sign [letter]. *Anesth Analg* 1994; 79: 192.

Supraclavicular brachial plexus block

Introduction

Supraclavicular block is performed above the clavicle where the brachial plexus transitions from trunks to divisions and is clustered into a small area adjacent to the subclavian artery.

Indications

Surgery to the forearm, elbow, wrist or hand (Figure 4.10). It reliably blocks the upper and middle trunks (median, radial, lateral and posterior cutaneous nerves of the forearm) but not the lower trunk (C8 and T1), so it is not a good choice for surgery on the medial side of the elbow, wrist, hand or little finger. The intercostobrachial nerve (T2) is missed in 90% of blocks; however, 10 ml of 1% lidocaine injected subcutaneously just distal to the axilla covers tourniquet pain in an awake patient.

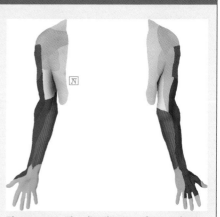

Figure 4.10 The distribution of anaesthesia following supraclavicular block. The area of anaesthesia is shown by the coloured areas, which correlate with Figure 6.5.

Specific contraindications

- Bilateral block – in case of accidental bilateral pneumothorax or phrenic nerve block.
- Bleeding diathesis – the subclavian artery is difficult to compress if accidently punctured.

Technique

Needle: 5 cm short bevelled insulated.

Landmark/nerve stimulator

The patient is slightly sat up, elbow flexed with forearm in lap and shoulder lowered to ensure the trunks lie above the clavicle.

The outer boundary of the dome of the pleura is noted (where the lateral border of the sternocleidomastoid meets the clavicle). The subclavian artery is palpated above the clavicle 25 mm lateral to this point. The needle is inserted 20 mm above the clavicle just behind the palpating finger, perpendicularly to the skin. After 2–5 mm, the needle is redirected caudally to pass under the palpating finger (aiming for the ipsilateral toe). The needle should not be inserted deeper than 25 mm, as the block takes place above the clavicle.

The upper trunk is usually encountered first (twitches of the deltoid), but the local anaesthetic may not spread to the lower trunk from this position,[1] so the needle is withdrawn and re-angled a few degrees posterior, but always parallel to the midline. The desired stimulation is of the anterior division of the middle trunk, or of the inferior trunk, producing hand contraction or finger extension.[2] 0.5 ml/kg (up to a maximum of 30 ml) of local anaesthetic is injected with frequent aspiration and low injection pressure.

Ultrasound-assisted block

A high-frequency probe is placed in the supraclavicular fossa and is directed caudally onto the first rib. The pulsatile subclavian artery is identified and the probe is rotated obliquely to obtain a cross-sectional view. The divisions of the brachial plexus are visualised lateral to the artery and superior to the first rib. A 50 mm needle is advanced in-plane into the space bordered by the rib inferiorly, the subclavian artery medially and the brachial plexus superolaterally. The nerves are tightly arranged in what appears to be a sheath formed by the prevertebral fascia that covers the scalene muscles. 25–30 ml of local anaesthetic is injected in this position.

a

Anterior   **Posterior**

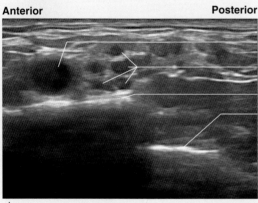

b

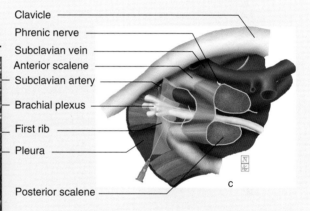

Clavicle
Phrenic nerve
Subclavian vein
Anterior scalene
Subclavian artery

Brachial plexus

First rib

Pleura

Posterior scalene

c

Figure 4.11 The anatomy of a supraclavicular block, with associated ultrasound image. The beam of the ultrasound is shown by the 'pane' on the anatomical figure. At this level the brachial plexus divides from trunks into divisions, as shown on both images. Note the proximity of the pleura, and hence the protection afforded by performing the block over the first rib.

Complications

- Phrenic nerve block.
- Pneumothorax – uncommon.
- Sympathetic nerve block and Horner's syndrome.
- Intravascular injection.
- Neuropraxia and neurologic injury – very uncommon.

References

1. Hickey R, Garland T, Ramamurthy S. Subclavian perivascular block: influence of location of paraesthesia. *Anesth Analg* 1989; **68**: 767–71.
2. Smith B. Distribution of evoked paraesthesia and effectiveness of brachial plexus block. *Anaesthesia* 1986; **41**: 1112–5.

Infraclavicular brachial plexus block

Introduction

Infraclavicular block is performed at the level of the cords of the brachial plexus.

Indications

Surgery to the forearm, wrist or hand (similar indications as for an axillary brachial plexus block).

Specific contraindications

- Coagulopathy is a relative contraindication, as compression of the artery is difficult at this level.

Technique

20–40 ml of local anaesthetic is required.

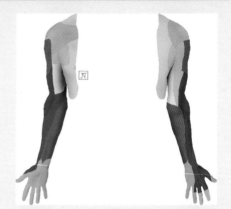

Figure 4.12 The distribution of anaesthesia following infraclavicular brachial plexus block. The area of anaesthesia is shown by the coloured areas, which correlate with Figure 6.5.

Landmark/nerve stimulator

Vertical infraclavicular approach

With the patient supine, a line is drawn between the jugular notch and the anterior process of the acromion and the midway point marked *beneath* the clavicle. This point should be lateral to the subclavian artery, which can be palpated *above* the clavicle. A 50 mm stimulator needle is inserted absolutely vertically from ventral to dorsal in the sagittal plane as close to the inferior surface of the clavicle as possible.

Posterior cord stimulation (elbow and wrist extension[1]) is associated with block success,[2] and is usually found at a depth of 2–4 cm.[3]

Stimulation of the lateral cord (forearm pronation and elbow flexion) suggests that the needle is too superficial, and of the medial cord (wrist flexion, intrinsic hand muscle contraction) that it is too medial. It is important to avoid medial angulation of the needle to avoid the pleura. If no twitches are found the needle is withdrawn and re-inserted, perpendicular in all planes, more medially initially and then more laterally.

Coracoid approach

The patient is supine and the coracoid process is palpated. The needle is inserted 2 cm inferior and 2 cm medial (1 cm in smaller patients) to the coracoid and directed from ventral to dorsal. The risk of pneumothorax increases as the needle insertion point moves more medially from the coracoid process. Posterior cord stimulation is associated with block success and is usually found at a depth of 4–8 cm.

Ultrasound-assisted block

Lateral sagittal approach

The patient is supine with the arm abducted to bring the artery and plexus closer to the skin.[4] A high-frequency probe is placed medial to the coracoid process, along a line from the acromioclavicular joint to the ipsilateral nipple. A transverse image of the axillary vessels is obtained (the artery appears superior to the vein), and the lateral, medial and posterior cords (named with respect to their position about the artery) are identified.

The needle is inserted in-plane or out-of-plane and slowly advanced towards the posterolateral border of the artery. A fascial 'click' may be felt, and local anaesthetic is injected in aliquots to achieve a U-shaped spread posterior to the artery.

The nerves may not be seen clearly because of their depth and the ultrasound artefact of posterior enhancement behind the artery, and the steep angle of approach makes visualisation of the needle difficult. Hydrolocation (visualising the needle tip by injecting a small amount of local anaesthetic) may be useful, but any injection that does not produce an identifiable hypoechoic bolus raises the risk of an intravascular injection.

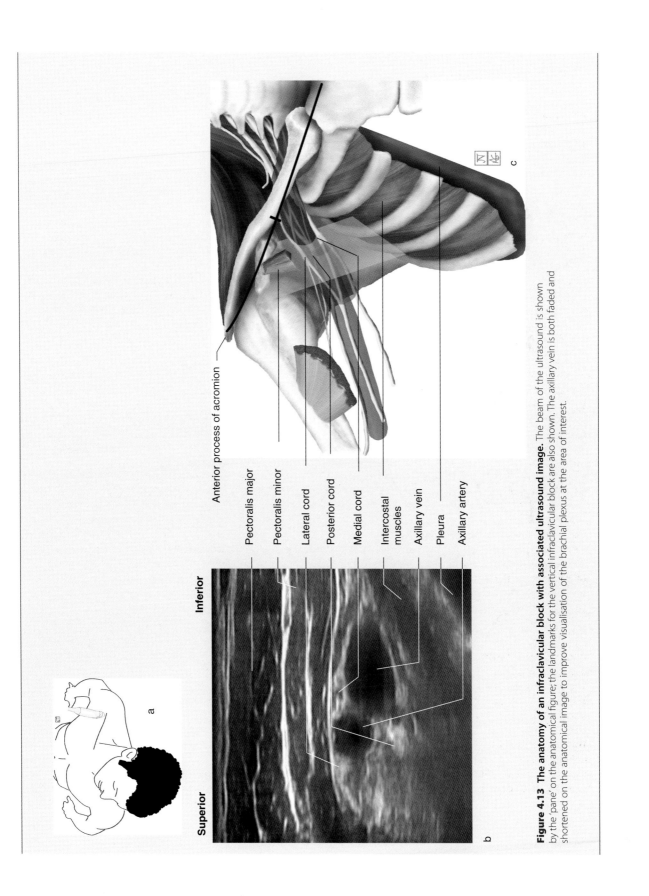

Superior

Inferior

Anterior process of acromion

Pectoralis major

Pectoralis minor

Lateral cord

Posterior cord

Medial cord

Intercostal muscles

Axillary vein

Pleura

Axillary artery

a

b

c

Figure 4.13 The anatomy of an infraclavicular block with associated ultrasound image. The beam of the ultrasound is shown by the 'pane' on the anatomical figure; the landmarks for the vertical infraclavicular block are also shown. The axillary vein is both faded and shortened on the anatomical image to improve visualisation of the brachial plexus at the area of interest.

Complications

- Pneumothorax.
- Vascular puncture and inadvertent intravascular injection.
- Phrenic nerve palsy.

Post-procedure checks

- Onset of the block is relatively slow. With nerve stimulator techniques, 60% are ready at 30 minutes, and with ultrasound 75% are ready at 20 minutes.[5]

References

1. Bowens C, Gupta R, O'Byrne W, *et al*. Selective local anesthetic placement using ultrasound guidance and neurostimulation for infraclavicular brachial plexus block. *Anesth Analg* 2010; **110**: 1480–5.
2. Lecamwasam H, Mayfield J, Rosow L, *et al*. Stimulation of the posterior cord predicts successful infraclavicular block. *Anesth Analg* 2006; **102**: 1564–8.
3. Koscielniak-Nielson Z, Rasmussen H, Hesselbjerg L, *et al*. Clinical evaluation of the lateral sagittal infraclavicular block developed by MRI studies. *Reg Anesth 30*: 329–34.
4. Koscielniak-Nielson Z, Frederiksen B, Rasmussen H, *et al*. A comparison of ultrasound-guided supraclavicular and infraclavicular blocks for upper extremity surgery. *Acta Anaesth Scand* 2009; **53**: 620–6.
5. Dingemans E, Williams S, Arcand G, *et al*. Neurostimulation in ultrasound-guided infraclavicular block: a prospective randomized trial. *Anesth Analg* 2007; **104**: 1275–80.

Suprascapular nerve block

Introduction

The suprascapular nerve is the main sensory nerve to the posterior and superior aspect of the shoulder joint (see text, section 2, above). Suprascapular nerve block provides analgesia to 70% of the shoulder joint,[1] but it does not provide cutaneous innervation and is inadequate as the sole block for awake surgery.

Indications

Suprascapular nerve block may be used for postoperative analgesia after shoulder arthroscopy,[2] for the management of chronic shoulder pain, and for the diagnosis of suprascapular neuropathy. It is an appropriate alternative to interscalene block for major subacromial surgery (such as rotator cuff repair) but provides inferior analgesia and has only a limited role in open shoulder surgery, where the anterior surgical approach is outside the region of suprascapular nerve innervation.[3]

Technique

Landmark/ nerve stimulator

The patient is in a sitting position with the head bent slightly forward. The midpoint of the scapular spine is marked. The needle entry point is 2 cm medial and 2 cm cranial from this point. A 50 mm stimulator needle is advanced in a lateral direction, at an angle of about 75° to the skin, towards the head of the humerus until bony contact is made with the floor of the fossa.[4] An alternative approach is to insert the needle perpendicular to the skin 1 cm superior to the midpoint of the scapula spine, aiming caudally until bony contact is made with the suprascapular fossa. After aspiration, 5–10 ml of local anaesthetic is injected to surround the nerve on the floor of the supraspinatus fossa.

If a nerve stimulator is used, stimulation of the suprascapular nerve produces external shoulder rotation and abduction.

Ultrasound-assisted block

The patient is in a sitting position. A high-frequency probe is positioned in the suprascapular fossa and tilted slightly anteriorly to view the inferior part of the suprascapular notch.

The suprascapular nerve is visualised on the floor of the suprascapular fossa beneath the fascia of supraspinatus.[5] Pulsation from the small suprascapular artery that lies adjacent to the nerve may be seen on the floor of the fossa. The needle is inserted in-plane or out-of-plane, and 5–10 ml of local anaesthetic elevates the fibres of the supraspinatus muscle from the suprascapular fossa and surrounds the suprascapular nerve.

Complications

- Pneumothorax (rare, < 1%).

References

1. Chan C, Peng P. Suprascapular nerve block: a narrative review. *Reg Anesth Pain Med* 2011; **36**: 358–73.
2. Ritchie E, Tong D, Chung F, *et al*. Suprascapular nerve block for postoperative pain relief in arthroscopic shoulder surgery: a new modality? *Anesth Analg* 1997; **84**: 1306–12.
3. Neal JM, McDonald SB, Larkin KL *et al*. Suprascapular nerve block prolongs analgesia after non-arthroscopic shoulder surgery, but does not improve outcome. *Anesth Analg* 2003; **96**: 982–6.
4. Meier G, Buettner J. *Peripheral Regional Anesthesia: an Atlas of Anatomy and Techniques*, 2nd edn. Stuttgart: Thieme, 2006; pp. 50–3.
5. Peng P, Wiley M, Liang J, *et al*. Ultrasound-guided suprascapular nerve block: a correlation with fluoroscopic and cadaveric findings. *Can J Anesth* 2010; **57**: 143–8.

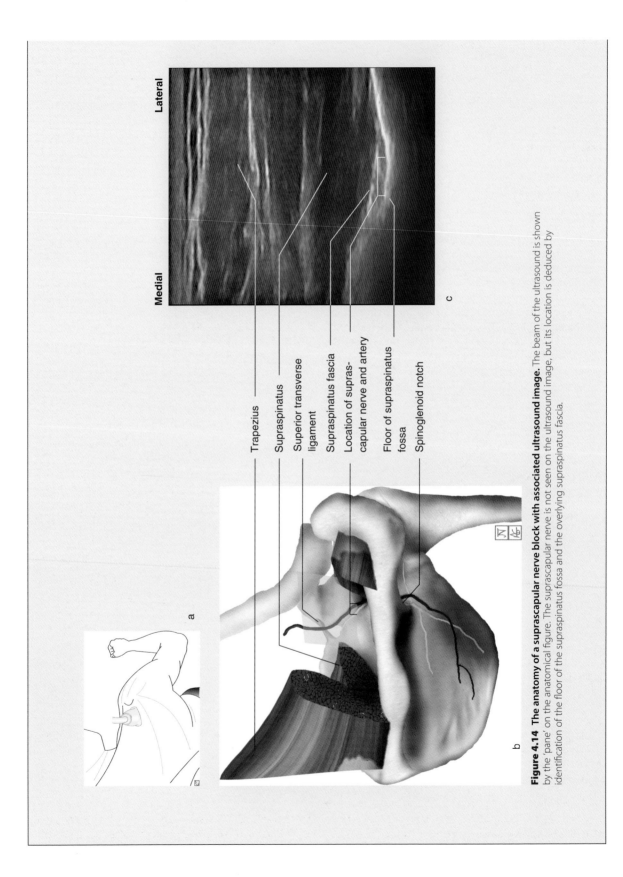

Lateral

Medial

Trapezius

Supraspinatus

Superior transverse
ligament

Supraspinatus fascia

Location of supras-
capular nerve and artery

Floor of supraspinatus
fossa

Spinoglenoid notch

a

b

c

Figure 4.14 The anatomy of a suprascapular nerve block with associated ultrasound image. The beam of the ultrasound is shown by the 'pane' on the anatomical figure. The suprascapular nerve is not seen on the ultrasound image, but its location is deduced by identification of the floor of the supraspinatus fossa and the overlying supraspinatus fascia.

3. The cervical sympathetic trunk

As in the thorax and abdomen, the cervical sympathetic trunk remains anterolateral to the vertebral column, lying on the deep prevertebral muscles of the neck.

The cervical sympathetic trunk receives no white rami communicantes. Fibres within the trunk originate from spinal levels T1–3 (occasionally T1–4 or 5) (see Chapter 1, section 6).

There are three sympathetic ganglia in the neck: the superior, middle and inferior cervical ganglia, found at levels C1–2, C6 and C7 respectively.

At the ganglia, ascending fibres synapse with the post-synaptic neurons which leave via grey rami communicantes to:

- Join spinal nerves.
- Form the cardiopulmonary splanchnic nerves. These are postganglionic neurons (unlike the splanchnic nerves of the abdomen) which pass along the vertebral bodies, trachea and oesophagus to reach the cardiac and pulmonary plexuses.
- Form plexuses around the vertebral and carotid arteries, travelling with the vessels to their destinations.

Specific sympathetic postsynaptic outputs from the ganglia are as follows:

i) Superior cervical ganglion

Via the ventral rami of C1–4.
On the left, the superficial cardiac plexus; on the right, the deep cardiac plexus.
The pulmonary plexus.

ii) Middle cervical ganglion

Via the ventral rami of C5–6.
The deep cardiac plexus.

iii) Inferior cervical ganglion

Fused with the first thoracic ganglion in 80% of people to form the stellate (or cervicothoracic) ganglion – an oval mass 2.5 cm long, 1 cm wide and 0.5 cm thick.

The stellate ganglion is found anterior to the transverse process of C7, just superior to the neck of the first rib on each side. It lies on the anterior surface of the longus colli muscle and may be above, within or beneath the prevertebral fascia (the literature is not clear which, but the stellate ganglion block seems to work best if placed beneath the fascia. This may be because it is the true location of the cervical sympathetic trunk, or because the distribution of local anaesthetic is preferential.)
Joins the ventral rami of C7–8 and enters the brachial plexus, and therefore supplies the arm.
Also supplies the deep cardiac plexus.

The cervical sympathetic trunk therefore supplies sympathetic innervation to the head and neck and in part to the arm and cardiopulmonary plexuses.

Stellate ganglion block

Introduction

Stellate ganglion block has a variable success rate,[1] and there are no high-quality placebo-controlled trials demonstrating long-term pain relief. This has contributed, along with serious side-effects, to a decline in its use.

Indications

Diagnosis and management of sympathetically mediated pain (such as phantom limb pain, post-herpetic neuralgia, cancer pain, refractory angina, orofacial pain and vascular headache), and of upper limb vascular insufficiency (Raynaud's syndrome, obliterative vascular disease, scleroderma and frostbite).[2]

Technique

Landmark

The patient is supine with the head turned away from the side to be blocked. The stellate ganglion is situated anterior to the transverse process of C7, just superior to the neck of the first rib on each side (Figure 4.3), but to avoid the pleura the block is often performed at the C6 level. The C6 level is located by palpating the cricoid cartilage and Chassaignac's tubercle (the anterior tubercle of the transverse process of C6). The sternocleidomastoid and carotid artery are retracted laterally, and the skin is pressed firmly onto Chassaignac's tubercle. The needle is advanced onto the tubercle, between the trachea and carotid sheath, and then redirected medially and inferiorly to contact the body of C6. The needle is withdrawn by 8-10 mm[3] to lie anterior to the belly of longus colli and beneath the prevertebral fascia, and radio-opaque dye is seen to spread craniocaudally between the tissue planes on anteroposterior and lateral image intensification.[4] Rapid dissipation of the dye suggests intravascular injection, and localised injection suggests intramuscular delivery.

After negative aspiration a 0.5 ml test dose containing 10-15 mcg adrenaline is administered to detect inadvertent injection into the vertebral artery. 10-15 ml of local anaesthetic is then injected in 3 ml aliquots, and the patient is placed in a sitting position to promote inferior spread to the stellate ganglion. Onset of Horner's syndrome indicates a successful block.

Ultrasound-assisted block

The patient is in a supine position with the head turned away from the side to be blocked. The neck is slightly extended by placing a pillow under the shoulders and the mouth slightly open to relieve tension in the neck. A high-frequency probe is used to obtain a transverse image of Chassaignac's tubercle (for guidance on image acquisition see Figure 4.9). The thyroid gland, carotid artery, compressible internal jugular vein and the oval-shaped structure of the longus colli muscle with overlying prevertebral fascia are identified (Figure 4.15). A parasagittal view may be useful to confirm the relationship of the vertebral body and disc, longus colli muscle and overlying fascia. The fascia is followed inferior to the transverse process of C6 to allow an in-plane needle approach from the lateral edge of the sternocleidomastoid. A 25G needle is inserted, avoiding the oesophagus, vertebral artery and inferior thyroid vessels, aiming posteriorly to a point beneath the prevertebral fascia but above the ventral surface of longus colli. Correct positioning of the needle tip is confirmed by visualisation of local anaesthetic expanding in the subfascial space. The injection is performed in aliquots after negative aspiration and after a test dose as described above.

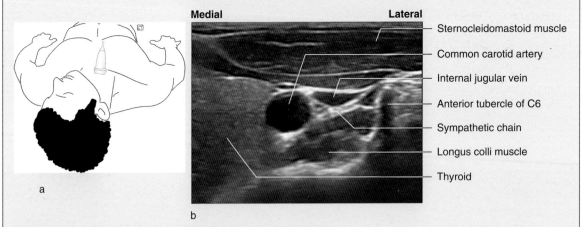

Medial — **Lateral**

- Sternocleidomastoid muscle
- Common carotid artery
- Internal jugular vein
- Anterior tubercle of C6
- Sympathetic chain
- Longus colli muscle
- Thyroid

a

b

Figure 4.15 Ultrasound image of the approach to a stellate ganglion block. For anatomical reference, refer to Figure 4.3.

Complications

- Intravascular injection.
- Neuraxial block.
- Nerve injury (vagus, brachial plexus, recurrent laryngeal nerve).
- Pneumothorax.
- Chylothorax.
- Oesophageal perforation.
- Haematoma (internal jugular vein puncture).

References

1. Forouzanfar T, Van Kleef M, Weber W. Radiofrequency lesions of the stellate ganglion in chronic pain syndromes: retrospective analysis of clinical efficacy in 86 patients. *Clin J Pain* 2000; **16**: 164–8.
2. Elias M . Cervical sympathetic and stellate ganglion blocks. *Pain Physician* 2000; **3**: 294–304.
3. Ates Y, Asik I, Ozgencil E, *et al*. Evaluation of the longus colli muscle in relation to stellate ganglion block. *Reg Anesth Pain Med* 2009; **34**: 219–23.
4. Erickson S, Hogan Q. CT-guided injection of the stellate ganglion: description of technique and efficacy of sympathetic blockade. *Radiology* 1993; **188**: 707–9.

The larynx

Found at vertebral levels C3–6 in the adult and C3–4 in the neonate.

1. Functions

- Protective sphincter of the respiratory tract.
- Phonation.
- Integral to the coughing mechanism.

2. Relations

i) Anterior

Pretracheal fascia, then investing layers of deep cervical fascia, superficial cervical fascia and platysma muscle, skin.

ii) Posterior

Laryngopharynx, middle and inferior pharyngeal constrictor muscles, prevertebral muscles, cervical vertebrae C3–6.

iii) Superior

Oropharynx.

iv) Inferior

Trachea.

v) Right and left lateral

Recurrent laryngeal nerves, internal and external laryngeal nerves, lobes of thyroid, superior thyroid artery and vein (and their branches, the superior laryngeal artery and vein).

3. Laryngeal skeleton

Consists of nine cartilages:

i) Three unpaired

Thyroid – between C4 and C5. Forms the laryngeal prominence (Adam's apple).
Cricoid – 'signet ring' shaped with band facing anteriorly, located at C6. The only complete ring of cartilage to encircle any part of the airway.
Epiglottic – attached to the thyroid cartilage by the thyro-epiglottic ligament.

ii) Three paired

Arytenoid.
Corniculate – provides attachments for some intrinsic laryngeal muscles.
Cuneiform – provides attachments for some intrinsic laryngeal muscles.

The hyoid bone is not part of the laryngeal skeleton.

4. Laryngeal ligaments

There are four extrinsic ligaments:
- Thyrohyoid membrane.
- Cricothyroid ligament.
- Cricotracheal ligament.
- Hyoepiglottic ligament.

The median cricothyroid ligament is the thickened anterior portion of the cricothyroid membrane. It is superficial and easily palpated between the thyroid and cricoid cartilages, making it the point of access

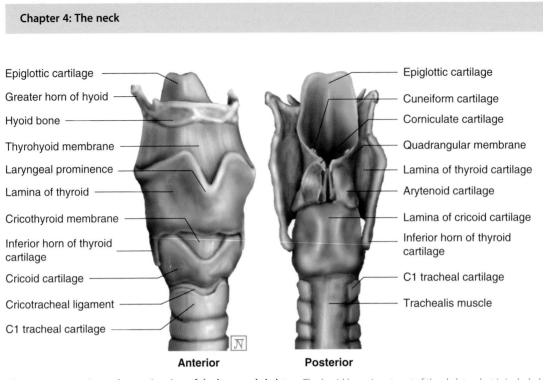

Epiglottic cartilage

Greater horn of hyoid

Hyoid bone

Thyrohyoid membrane

Laryngeal prominence

Lamina of thyroid

Cricothyroid membrane

Inferior horn of thyroid cartilage

Cricoid cartilage

Cricotracheal ligament

C1 tracheal cartilage

Epiglottic cartilage

Cuneiform cartilage

Corniculate cartilage

Quadrangular membrane

Lamina of thyroid cartilage

Arytenoid cartilage

Lamina of cricoid cartilage

Inferior horn of thyroid cartilage

C1 tracheal cartilage

Trachealis muscle

Anterior   **Posterior**

Figure 4.16 Anterior and posterior view of the laryngeal skeleton. The hyoid bone is not part of the skeleton but is included as it is palpable clinically. The cricothyroid membrane is thickened anteriorly and medially to form the median cricothyroid ligament. The vocal ligaments are seen between the arytenoids on the posterior view.

in the 'can't intubate, can't ventilate' airway situation (see box on *Surgical and cannula cricothyroidotomy*, below).

The lateral cricothyroid ligament extends laterally and posteriorly between the thyroid and arytenoid cartilages. Its free edge forms the vocal ligament, which when covered in mucous membrane is known as the vocal fold or true vocal cord.

Between the epiglottic and arytenoid cartilages is the quadrangular membrane, the free inferior edge of which is the vestibular ligament. When covered in mucous membrane, this is known as the vestibular fold or false vocal cord. The free superior edge is the aryepiglottic ligament, the skeleton of the aryepiglottic fold, which is visible on laryngoscopy (Figure 4.17).

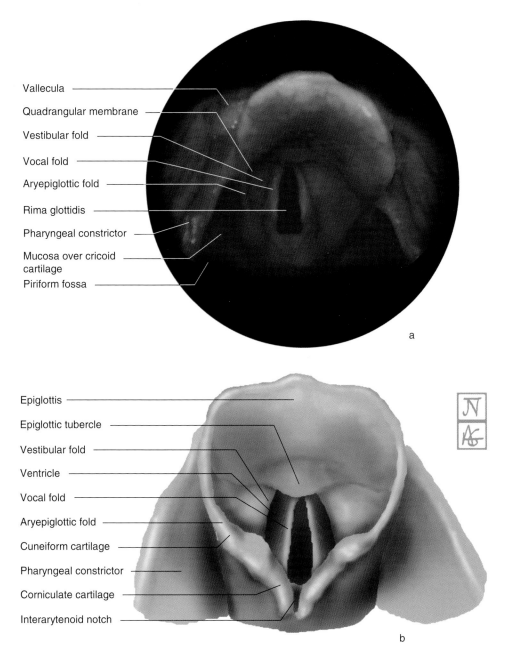

Vallecula

Quadrangular membrane

Vestibular fold

Vocal fold

Aryepiglottic fold

Rima glottidis

Pharyngeal constrictor

Mucosa over cricoid cartilage

Piriform fossa

a

Epiglottis

Epiglottic tubercle

Vestibular fold

Ventricle

Vocal fold

Aryepiglottic fold

Cuneiform cartilage

Pharyngeal constrictor

Corniculate cartilage

Interarytenoid notch

b

Figure 4.17 Internal views of the larynx. (a) Laryngoscopic view and (b) anatomical image (of different elevation) to aid structure identification. The quadrangular membrane forms the lateral wall of the vestibule (the space superior to the vestibular folds). The ventricle is the space between the vestibular and vocal folds. The rima glottidis is the space between the vocal folds. The vestibular folds are also known as the false vocal cords.

Surgical and cannula cricothyroidotomy

Introduction

The situation in which neither intubation nor ventilation is possible is commonly called 'can't intubate; can't ventilate' (CICV).[1] Rescue techniques involve accessing the airway through the cricothyroid membrane, which is approximately 10 mm high and 11 mm wide (Figure 4.16)[2] using one of three techniques:
- Narrow-bore cannula: internal diameter (ID) < 2 mm.
- Wide-bore cannula: ID > 4 mm.
- Surgical: scalpel and size 6.0 ID tube.

Indications

CICV – readers are referred to the Difficult Airway Society (DAS) guidelines.[1] Risk factors for CICV include obesity,[3] upper airway or midfacial trauma (including inhalational or thermal), airway oedema, infection, tumour and foreign body.

Some authors advocate prophylactic placement of a transtracheal catheter before inducing anaesthesia, when a cricothyroidotomy might be required.[4]

Specific contraindications

There are no contraindications in a life-threatening airway emergency, but the techniques are made more difficult in the presence of a large laryngeal tumour, obesity, neck pathology or coagulopathy.[5]

Technique

Failure to identify the cricothyroid membrane is common,[6] and confirmation of its position prior to induction of anaesthesia is recommended. Ultrasound may be used to identify the point of puncture and to avoid blood vessels.

Landmark

All techniques require head and neck extension, which may be facilitated with a bag of fluid or a bolster under the scapulae and removal of the pillow.

Narrow-bore cricothyroidotomy

The cricoid cartilage is immobilised and the cricothyroid membrane punctured in the midline with a kink-resistant cannula (ID < 2 mm, e.g. VBM Ravussin cannula) attached to a 5–10 ml syringe partially filled with saline. Correct needle placement is confirmed by aspiration of air through the saline (and capnography when the catheter is placed prophylactically). The cannula is advanced into the trachea and the needle is removed. The cannula is stabilised by an assistant, a high-pressure oxygen source is attached, and a low initial driving pressure is used (< 1 bar). The chest should be seen to rise and then fall when insufflation ceases. Exhalation mandates a patent upper airway, and an oral or supraglottic airway may be required.

Wide-bore cricothyroidotomy

A wide-bore cannula (ID > 4 mm) with a cuffed tube allows conventional ventilation with a standard anaesthetic breathing system.
- The Quicktrach II (VBM) is a wide-bore cuffed cannula mounted on a trochar. The cricoid cartilage is immobilised and the cricothyroid membrane punctured in the midline at 90° to the skin. Once air is aspirated, the angle of insertion is lowered to 60° and the device is advanced up to the red detachable stopper, which is removed prior to advancing the cannula over the needle. The cuff is inflated, and a conventional breathing system is connected.
- The Melker Universal Cricothyrotomy Catheter Set (Cook Medical) uses a Seldinger technique. The cricothyroid membrane is stabilised, a vertical midline incision is made, and the supplied needle (attached to a syringe partially filled with saline) is advanced through the incision at 45° in a caudal direction. Following aspiration of air and passage of the guidewire through the needle, the needle is removed and the airway catheter/dilator is passed over the guidewire until it appears through the handle of the assembly. The guidewire is held in place and the airway catheter is advanced into the trachea. The guidewire and dilator are removed simultaneously. The cuff is inflated and a conventional breathing system is connected.

Surgical cricothyroidotomy

The cricoid cartilage is immobilised between the thumb and middle finger of the non-dominant hand. A horizontal incision is made through the lower part of the cricothyroid membrane.[7] An initial vertical incision may be necessary in obese patients to identify the location of the cricothyroid membrane. Once the blade is in the trachea either a tracheal hook or a Trousseau (tracheal) dilator is inserted before the blade is removed. The incision is enlarged with the dilator. A bougie, loaded with a size 6.0 tracheal tube, is then introduced through the incision, aiming caudally into the trachea. Rotating the dilator (thereby elevating the cricoid cartilage) may facilitate easier passage of the tube into the trachea. The bougie is removed, the tracheal tube is connected to a standard breathing system, and its position is confirmed with capnography.

Complications

Cannula techniques have been associated with a significant failure rate (narrow-bore 63%, wide-bore 43%). As the reasons for this are multifactorial, the optimal technique is less important than preparedness and the overall management of the situation.

- Pneumothorax (rare, < 1%).
- False passage and surgical emphysema.
- Bleeding.
- Failure (preferable to not attempting in the CICV situation).
- Death.

Post-procedure checks

- Correct tracheal placement of the device is confirmed by chest rise and fall with ventilation and with capnography. No more than one or two breaths should be delivered while assessing the success of the procedure, as ventilation through any misplaced device is likely to produce significant surgical emphysema, further hampering subsequent attempts at rescuing the airway.

References

1. Henderson J, Popat M, Latto P, *et al.* Difficult Airway Society guidelines for management of the unanticipated difficult intubation. *Anaesthesia* 2004; **59**: 675–94.
2. Dover K, Howdieshell T, Colborn G. The dimensions and vascular anatomy of the cricothyroid membrane: relevance to emergent surgical airway access. *Clin Anat* 1996; **9**: 291–5.
3. Cook T, Woodall N, Frerk C, Fourth National Audit Project. Major complications of airway management in the UK: results of the Fourth National Audit Project of the Royal College of Anaesthetists and the Difficult Airway Society. *Br J Anaesth* 2011; **106**: 617–31.
4. Geriq H, Schnider T, Heidegger T. Prophylactic percutaneous transtracheal catheterisation in the management of patients with anticipated difficult airways: a case series. *Anaesthesia* 2005; **60**: 801–5.
5. Hamaekers A, Henderson J. Equipment and strategies for emergency tracheal access in the adult patient. *Anaesthesia* 2011; **66**: 65–80.
6. Elliott D, Baker P, Scott M, *et al.* Accuracy of surface landmark identification for cannula cricothyroidotomy. *Anaesthesia* 2010; **65**: 889–94.
7. Airway and ventilatory management. In *Advanced Trauma Life Support for Doctors. Student Course Manual*, 6th edn. Chicago, IL: American College of Surgeons Committee on Trauma, 1997; 83–5.

5. Laryngeal cavity

The laryngeal cavity is divided into:
- Vestibule – superior to the vestibular folds (continuous with the laryngopharynx).
- Ventricle – between the vestibular and vocal folds.
- Rima glottidis – between the vocal folds.
- Infraglottic cavity – between the vocal folds and the cricoid cartilage (continuous with the trachea).

The glottis refers to vocal folds and the space between them (the rima glottidis).

Viewing the larynx requires alignment of the mouth, oropharynx and larynx. This requires flexion of the neck to align the oropharynx and larynx, and extension of the head at the atlanto-occipital joint to bring the mouth into alignment. This position has been described as 'sniffing the morning air'.

6. Laryngeal muscles

Consist of three extrinsic and six intrinsic muscles.

i) Extrinsic

Move the larynx as a whole.
- Sternothyroid – depresses larynx.
- Thyrohyoid – elevates larynx.
- Inferior constrictor of the pharynx – constricts pharynx.

ii) Intrinsic

Summarised in Table 4.1.

7. Neurovascular supply

i) Arterial

External carotid → superior thyroid → superior laryngeal artery.
Subclavian → inferior thyroid → inferior laryngeal artery.

ii) Venous

Corresponding superior and inferior thyroid veins, which drain to the internal jugular and left brachiocephalic veins respectively.

iii) Nervous

Refer to Figure 3.21.
The anterior surface of the epiglottis is innervated either by the glossopharyngeal nerve or the vagus nerve, or by a combination of the two.
All remaining innervation is derived from the vagus nerve:
1) The superior laryngeal nerve – passes deep to the internal and external carotid arteries and divides into:
   - Internal laryngeal nerve – sensory to the mucosa superior to the vocal folds.
   - External laryngeal nerve – motor to cricothyroid muscle.
2) The recurrent laryngeal nerve – sensory to the mucosa inferior to the vocal fold and motor to all intrinsic muscles of the larynx except cricothyroid.
   Anaesthesia of these nerves and therefore of the larynx is discussed in the box

Table 4.1 Intrinsic muscles of the larynx.

Muscle	Action on vocal cords	Innervation
Cricothyroid	Tensor	External laryngeal nerve
Posterior cricoarytenoid	Abductor	Recurrent laryngeal nerve
Lateral cricoarytenoid	Adductor	
Thyroarytenoid	Relaxor	
Transverse and oblique arytenoid	Closes rima glottidis	
Vocalis	Posterior relaxor, anterior tensor	

on *Regional anaesthesia for awake fibreoptic intubation* (see Chapter 3, The nose).

The recurrent laryngeal nerves branch off the vagus nerve – on the right as it passes anterior to the subclavian artery and on the left as it passes over the arch of the aorta. The recurrent laryngeal nerves loop round these vessels before ascending in the tracheo-oesophageal groove to reach the larynx.

The recurrent laryngeal nerves may therefore be damaged in both the thorax (e.g. oesophageal malignancy, aneurysm of the aortic arch) and the neck (e.g. thyroid surgery). Unilateral damage results in paralysis of the corresponding vocal cord, which lies near the midline, manifesting as hoarseness. Bilateral damage results in complete loss of vocal power and a flap valve effect as the cords flutter together on inspiration, manifesting as stridor and severe dyspnoea.

Percutaneous tracheostomy

Introduction

Percutaneous dilational tracheostomy is a commonly performed procedure on the ICU. It is a procedure performed electively; it is not a means of emergency access to the airway (see box on *Surgical and cannula cricothyroidotomy*, above).

Indications

Maintain the airway.
Protect the airway.
Facilitate weaning from mechanical ventilation.
Access to the airway for bronchial toilet.

Specific contraindications

Relative:
- Grossly abnormal anatomy.
- Coagulopathy.
- Cervical spine injury.
- Large vessels identified superficial to the trachea at site of puncture (e.g. left brachiocephalic vein). Consider surgical tracheostomy under these circumstances.

Pre-procedure checks

Ensure that the patient's clotting status is not deranged, and that drugs which may render the patient coagulopathic are withheld or stopped.

The procedure requires an operator and an airway specialist as well as adequate support personnel. It is the responsibility of the airway specialist to provide general anaesthesia and to maintain the airway at all times. It is therefore essential that a laryngoscope and a new tracheal tube are immediately to hand and a difficult airway trolley is present in the immediate vicinity of the patient's bed space.

A bolster is placed behind the scapulae to facilitate extension of the neck.

The tracheal tube is withdrawn to the glottis and pressure-controlled ventilation of the lungs with 100% oxygen is confirmed.

Following skin preparation with 2% chlorhexidine gluconate in 70% isopropyl alcohol, local anaesthetic with 1:200,000 adrenaline is infiltrated at the point of skin puncture and adequate time for vasoconstriction is allowed.

Technique

A variety of percutaneous tracheostomy kits are available. The underlying principle is that of a Seldinger technique. The equipment is checked, and hydrophilic items are soaked with saline to ensure smooth passage through the skin.

Landmark

A syringe partially filled with saline is mounted on the cannula. The trachea is gripped with the thumb and index finger of the non-dominant hand and the trachea is punctured between the first and second or the second and third tracheal rings while aspirating continuously on the syringe. Positive aspiration of air through the saline and fibreoptic visualisation of the cannula in the trachea, via the tracheal tube, confirms tracheal placement of the cannula. The cannula is slid off the needle and a wire passed. Inferior passage of the wire into the trachea is again confirmed fibreoptically. Sequential dilation under fibreoptic guidance then occurs; a horizontal cut may be made in the skin to facilitate this but is not always necessary. Depth markers on the wire should be visible at all times, to prevent it from being inadvertently forced into the bronchial tree during the dilational process. The tracheostomy tube (mounted on a dilator) is then passed over the wire into the trachea and its position confirmed by fibreoptic examination of the trachea via the tracheostomy tube. The cuff of the tracheostomy tube is then inflated. Only when the trachea is visualised through the tracheostomy tube is ventilation through it commenced. Capnography and flow patterns on the ventilator provide additional confirmation, prior to securing the tracheostomy tube and removing the orotracheal tube.

Ultrasound-assisted approach

Ultrasound may be used to examine the trachea prior to the procedure, with the primary aim of assessing the point of puncture and identifying vessels which would make bleeding a significant risk. Experienced users may perform the procedure under ultrasound guidance.

Complications

Early:
- Loss of the airway and death.
- Bleeding.
- Pneumothorax.
- Infection.
- Oesophageal injury.

Late:
- Bleeding.
- Tracheal stenosis.
- Tracheo-oesophageal fistula.

Post-procedure checks

- Chest x-ray to exclude pneumothorax and confirm position of the tube.
- Plans for airway management in the event of inadvertent tracheostomy displacement should be made in advance and appropriate algorithms should be immediately available.[1,2]

References

1. National Tracheostomy Safety Project. 2013. http://www.tracheostomy.org.uk (accessed November 2013).
2. New-2-ICU. Displaced tracheostomy algorithms 1 and 2. 2013. http://www.new2icu.co.uk (accessed November 2013).

The thyroid gland

Consists of two lobes lying anterolateral to the trachea between C5 and T1.

The lobes are united by an isthmus, usually found anterior to the second or third tracheal rings.

Contained within pretracheal fascia.

Enlargement of the thyroid gland may compress the trachea and oesophagus and may damage the recurrent laryngeal nerves. Extensive enlargement may result in retrosternal extension of the gland, which may cause distal compression of the trachea. Following anaesthesia, any compression may be exacerbated by the loss of negative intrapleural pressure. If the compression is distal, it may be difficult to pass an endotracheal tube beyond the obstruction, warranting turning the patient onto his or her side or front to gain the assistance of gravity. A detailed history and relevant investigations (such as flow volume loops, CT or MRI) is therefore essential prior to anaesthesia of patients with enlarged thyroid glands.

The thorax

The thoracic wall

1. Thoracic boundaries

The thorax is bounded by:

i) Thoracic inlet

The first thoracic vertebra.
The first pair of ribs and their respective cartilages.
The superior border of the manubrium.

ii) Thoracic outlet

The twelfth thoracic vertebra.
The eleventh and twelfth pairs of ribs.
The costal cartilages of ribs 7–10.
The xiphisternal joint.
The aperture is closed by the musculotendinous diaphragm.

iii) Thoracic wall

Formed by the ribs and intercostal muscles, vertebral bodies and intervertebral discs of T1–12.

The thoracic skeleton, which forms part of the thoracic wall, comprises:

- Twelve thoracic vertebrae and intervertebral discs (more detail in Chapter 2).
- Twelve pairs of ribs and costal cartilages.
- The sternum.

2. Typical ribs

Typical ribs (3–9) have the following features (Figure 5.1):

- A head (has two facets for articulation with the facet of the corresponding vertebra and that of the vertebra above).
- A neck.
- A tubercle (for articulation with the corresponding vertebral transverse process).
- A shaft (has an angle, and a costal groove which contains the neurovascular bundle).

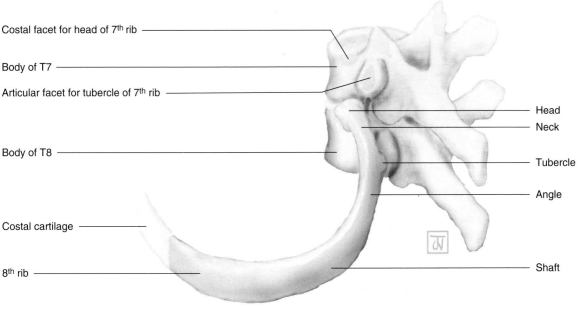

Costal facet for head of 7th rib

Body of T7

Articular facet for tubercle of 7th rib

Body of T8

Costal cartilage

8th rib

Head

Neck

Tubercle

Angle

Shaft

Figure 5.1 Posterolateral view of a typical (eighth) rib.

3. Atypical ribs

Atypical ribs (ribs 1, 2, 10, 11 and 12) have features which differentiate them from the typical ribs:

i) 1st rib

Shortest, flattest and has the greatest curvature of all ribs. Fracture of the first rib is indicative of extreme traumatic force and should alert the clinician to the possibility of other serious injuries. Has a head with one articular facet for articulation with the body of T1 vertebra, a neck, a tubercle for articulation with the transverse process of T1 vertebra, and an angle.

Features, from posterior to anterior (Figure 5.2):
- Insertion of middle scalene muscle.
- Groove for subclavian artery and posterior trunk of the brachial plexus.
- Scalene anterior tubercle for attachment of anterior scalene muscle.
- Groove for subclavian vein.

Serratus anterior inserts laterally, subclavius to the superior portion anteriorly, and the intercostal muscles to the lateral margin.

ii) 2nd rib

Thinner and longer than the first rib, with a tubercle for muscle attachment.

iii) 10th, 11th, 12th ribs

Have only one facet on their heads.
Ribs 11 and 12 have no necks or tubercles.

1st Rib

Associated anatomy

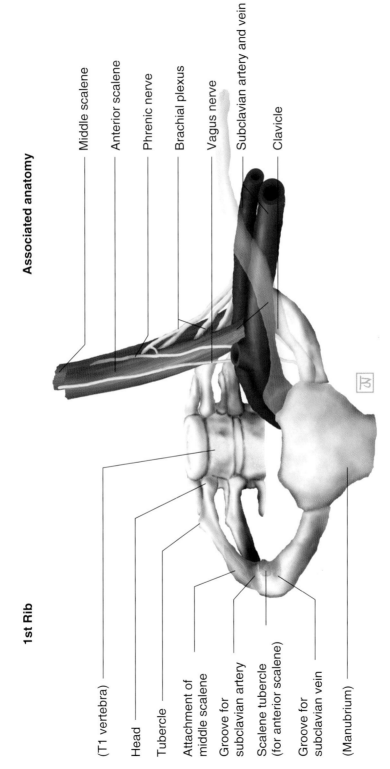

Middle scalene

Anterior scalene

Phrenic nerve

Brachial plexus

Vagus nerve

Subclavian artery and vein

Clavicle

(T1 vertebra)

Head

Tubercle

Attachment of
middle scalene

Groove for
subclavian artery

Scalene tubercle
(for anterior scalene)

Groove for
subclavian vein

(Manubrium)

Figure 5.2 First rib and associated anatomy. The vagus nerve lies within the carotid sheath, which has been omitted for illustrative purposes.

4. The clavicle

The medial two-thirds is convex anteriorly, the lateral one-third is concave.

Acts as a strut to the upper limb and protects the subclavian vessels and brachial plexus which lie beneath.

Relations:

- The subclavian vein is anterior and inferior to the subclavian artery.
- The phrenic nerve lies between the subclavian vessels.

- Posterior and lateral to the subclavian artery lies the brachial plexus. The superior, middle and inferior trunks of the brachial plexus each divide into anterior and posterior divisions at the level of the clavicle.
- The surface landmark of the plexus is a point midway between the anterior process of the acromion and the jugular notch.

Subclavian venous access

Introduction

The subclavian vein runs deep to the medial third of the clavicle (Figure 5.3).

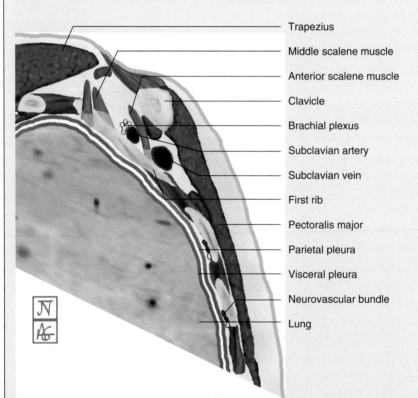

Trapezius

Middle scalene muscle

Anterior scalene muscle

Clavicle

Brachial plexus

Subclavian artery

Subclavian vein

First rib

Pectoralis major

Parietal pleura

Visceral pleura

Neurovascular bundle

Lung

Figure 5.3 The anatomy relevant for subclavian line insertion. Sagittal section at the lateral margin of the first rib, showing the anatomical relations of the clavicle.

Indications

See box on *Internal jugular venous access*.

Specific contraindications

See box on *Internal jugular venous access*.
- The subclavian artery is non-compressible, and inadvertent puncture may produce life-threatening haemorrhage, particularly in the face of coagulopathy.

Pre-procedure checks

See box on *Internal jugular venous access*.

Technique

Aseptic Seldinger technique – see box on *Internal jugular venous access*.

Landmark

The needle is inserted 1 cm inferior to the point of union of the medial two-thirds and lateral third of the clavicle. A finger is placed in the jugular notch to identify the needle direction and angle of approach to be achieved. The needle angle is flattened once the tip is below the clavicle, as the greater the vertical intent the higher the risk of pneumothorax or arterial puncture.

Ultrasound-assisted block

A high-frequency probe is placed just medial to the coracoid process and just below the clavicle in a parasagittal plane. The axillary artery and vein are identified deep to the pectoralis major and minor muscles. The probe is moved medially to follow the vessels as they run deep to the clavicle. The subclavian vein lies inferior and anterior to the artery, and the lung can be seen directly posterior to the vessels. The probe is centred over the subclavian vein, and the needle is introduced out-of-plane.

Complications

See box on *Internal jugular venous access*.
- Pneumothorax (1–3%). The subclavian approach carries the greatest risk of pneumothorax/haemothorax and is most likely to advance into an incorrect position (ascending the internal jugular vein or crossing into the opposite subclavian vein). It does however have the lowest incidence of infection and arterial puncture.
- Subclavian stenosis.

Post-procedure checks

See box on *Internal jugular venous access*.

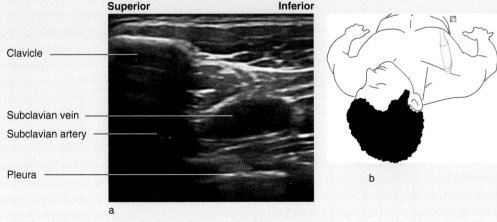

Figure 5.4 The ultrasound approach to the subclavian vein.

5. Intercostal muscles

From external to internal:

i) External intercostal

Eleven pairs, running from the lower border of the rib above to the upper border of the rib below. The fibre direction is inferomedial ('hands in pockets'). Become more fibrous anteriorly, forming the external intercostal membrane. Continuous inferiorly with external oblique muscles of the abdominal wall.

ii) Internal intercostal

Eleven pairs, running at right angles to the external intercostals.
Become more fibrous posteriorly, forming the internal intercostal membrane at the angle of the rib.
Continuous inferiorly with internal oblique muscles of the abdominal wall.

iii) Innermost intercostal

Similar and deep to internal intercostal muscles – pass between internal surfaces of adjacent ribs.

iv) Subcostal and transversus thoracis

Cross more than one intercostal space, so are not strictly intercostal muscles.
Subcostal (found posteriorly) run in the same direction as the internal intercostals.
Transversus thoracis attaches to the xiphoid process and is continuous with transverse abdominal muscle of the abdominal wall.

6. Neurovascular supply

Vessels and nerves lie between internal and innermost intercostal muscles in the costal groove.
From superior to inferior (mnemonic – VAN):

i) Venous

Eleven intercostal veins and one subcostal vein (beneath the twelfth rib) on each side.
Drain anteriorly to the internal thoracic veins; posteriorly to the azygous or hemiazygous system.

ii) Arterial

Each intercostal space is supplied by a large posterior intercostal artery and a small pair of anterior intercostal arteries.
The posterior intercostal arteries between T1 and T2 are supplied by the superior intercostal artery (arising from the subclavian artery), and between T3 and T12 they arise directly from the thoracic aorta.
The anterior intercostal arteries (T1–9) are supplied by the internal thoracic arteries (arise from the subclavian artery).

iii) Nervous

Ventral rami of T1–12.
T7–12 continue between internal oblique and transverse abdominal muscles (the transverse abdominal plane) to supply the abdominal wall.
Branches:
- Rami communicantes – to and from the sympathetic trunk.
- Collateral – to intercostal muscles, pleura and rib periosteum.
- Cutaneous – to skin. These branches emerge at the anterior axillary line, and divide into anterior and posterior branches (Figure 5.6). This has implications for the site of an intercostal nerve block (see box on *Intercostal nerve block*, below).
- Muscular – to intercostal, levatores costarum and serratus posterior muscles.

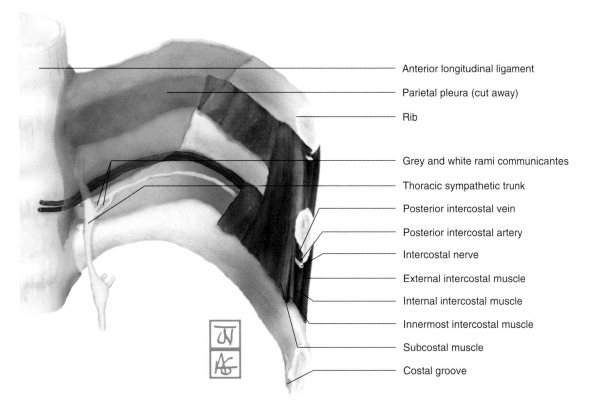

Anterior longitudinal ligament

Parietal pleura (cut away)

Rib

Grey and white rami communicantes

Thoracic sympathetic trunk

Posterior intercostal vein

Posterior intercostal artery

Intercostal nerve

External intercostal muscle

Internal intercostal muscle

Innermost intercostal muscle

Subcostal muscle

Costal groove

Figure 5.5 Posterior intercostal space with associated muscles and neurovascular bundle.

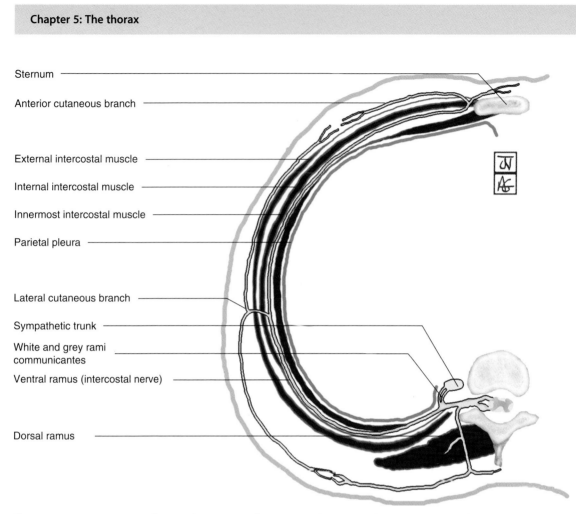

Sternum

Anterior cutaneous branch

External intercostal muscle

Internal intercostal muscle

Innermost intercostal muscle

Parietal pleura

Lateral cutaneous branch

Sympathetic trunk

White and grey rami
communicantes

Ventral ramus (intercostal nerve)

Dorsal ramus

Figure 5.6 Transverse section showing the anatomy of an intercostal nerve. Note how the nerve passes between internal and innermost intercostal muscles. Note also the position of the lateral cutaneous branch, explaining why an intercostal block must be placed posterior to the mid-axillary line to be effective.

Intercostal nerve block (ICNB)

Introduction
Provides anaesthesia to the skin and musculature of the chest and abdominal wall. The technique provides post-operative analgesia, is opioid-sparing, and allows improved pulmonary mechanics in the conditions listed below.

Indications
Analgesia for rib fractures and postsurgical pain after thoracotomy, mastectomy or cholecystectomy. Blockade of two levels above and below the level of surgical incision is needed.

Specific contraindications
- Where pneumothorax would be disastrous.
- ICNB above T7 may be difficult because of the scapulae.

Technique
Needle: 4–5 cm 20–22G short-bevelled needle.

Landmark

The patient may be prone, sitting, or in the lateral position with the arms positioned so that the scapulae are pulled laterally. The block may be performed anywhere posterior to the mid-axillary line, prior to the origin of the lateral cutaneous branch (which innervates adjacent skin – see text and Figure 5.6). It is commonly performed 6–8 cm lateral to the spinous processes (the angle of the rib) where the rib is easy to palpate and the costal groove is theoretically the widest – reducing the chance of pleural puncture. Medial to the angle of the rib there is little tissue between the nerve and the parietal pleura. The neurovascular bundle lies under the inferior aspect of the rib in the costal groove. The skin is infiltrated with a small volume of lidocaine. The skin is pulled cranially, and with the needle angled 20° cephalad contact is made with the rib (usually at a depth of < 1 cm). The skin is then relaxed and the needle advanced 3 mm at the same angle until a subtle give or 'pop' is felt. The average distance to the pleura is about 8 mm. 3 ml of local anaesthetic spreads 4–6 cm along the costal groove. 20 ml may spread to the paravertebral space.

Ultrasound-assisted block

The chest wall is imaged with a high-frequency probe placed vertically in a parasagittal plane and the intercostal space is visualised. The hypoechoic ribs are identified with the intercostal muscles between them. Deep to the intercostal muscles are the pleura and lungs, with the neurovascular bundle lying between the innermost intercostal and the internal intercostal muscles (Figure 5.5). The nerves may then be blocked using an in-plane or out-of-plane approach.

Complications

- Pneumothorax (< 1%) and lung injury (rare).
- Toxicity – rapid absorption with peak arterial plasma concentration in under 5–10 min.
- Spinal anaesthesia (rare).

Post-procedure checks

- Post-procedure chest x-ray if concerned about pneumothorax.

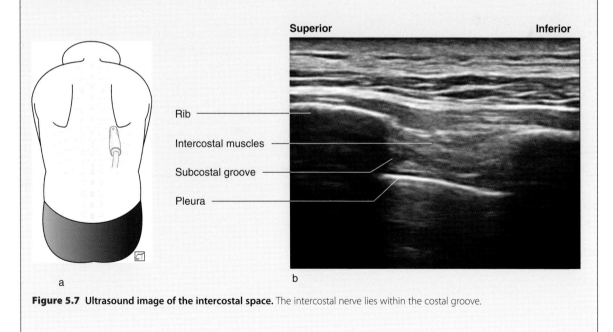

Superior　　　　　　　　　　　　　　　　　　　　**Inferior**

Rib

Intercostal muscles

Subcostal groove

Pleura

a　　　　　　　　　　　　　　　　　　　　b

Figure 5.7 Ultrasound image of the intercostal space. The intercostal nerve lies within the costal groove.

Lateral cutaneous branch of intercostal nerve

External intercostal muscle

Internal intercostal muscle (external intercostal muscle removed)

Innermost intercostal muscle (both external and internal intercostal muscles removed)

10th intercostal nerve beneath 10th rib

Reflected internal intercostal muscle

External oblique muscle

Internal oblique muscle

Transverse abdominal muscle with T11 nerve running in transverse abdominal plane

Posterior rectus sheath

Rectus abdominis muscle

Anterior rectus sheath

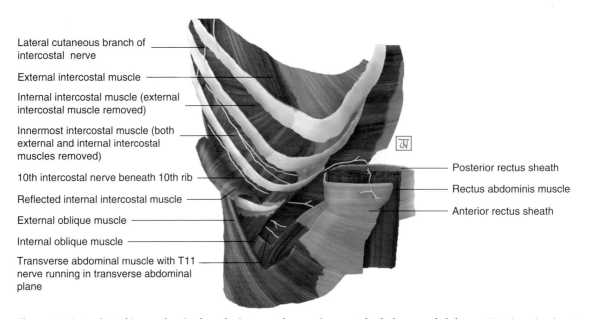

Figure 5.8 Anterolateral image showing how the intercostal nerves innervate both thorax and abdomen. Note how the plane in which the nerves travel is consistent between thorax and abdomen, becoming the transverse abdominal plane (TAP) in the abdomen. Note that the nerves enter the rectus sheath posteriorly, explaining the rationale for depositing local anaesthetic between the rectus muscle and the posterior rectus sheath during a rectus sheath block.

7. Thoracic paravertebral space

A potential wedge-shaped space found between the thoracic vertebrae and the parietal pleura (Figure 5.9).

i) Boundaries

- Medial – vertebral bodies and pedicles, intervertebral foramina and intervertebral discs. Communication with the epidural space is via the intervertebral foramina.
- Lateral and anterior – parietal pleura in the thoracic region.
- Posterior – transverse processes of thoracic vertebrae, heads and necks of the ribs and the superior costotransverse ligaments.
- Superior/inferior – T1/T12 levels.

ii) Contents

- Loose fat and areolar tissue.
- Spinal nerves T1–12.
- Sympathetic trunk, grey and white rami communicantes (see Chapter 1, section 6).
- Blood vessels.

Local anaesthetic agents deposited within this space will spread over two to four spinal segments, providing unilateral analgesia and limited sympathetic blockade, without the risks associated with spinal or epidural blockade.

Note: paravertebral spaces are found at all but the sacral and coccygeal levels. However, paravertebral spaces outside the thoracic region do not communicate freely, limiting the spread of local anaesthetic solutions and therefore the usefulness of the technique. Remember, however, that the thoracic dermatomes extend over the chest and most of the abdomen (Figure 5.8).

Lung and visceral pleura

Parietal pleura

Sympathetic trunk

White and grey rami
communicantes

Ventral and dorsal roots

Spinal nerve

Paravertebral space

Superior costotransverse
ligament

Rib

Intervertebral foramen

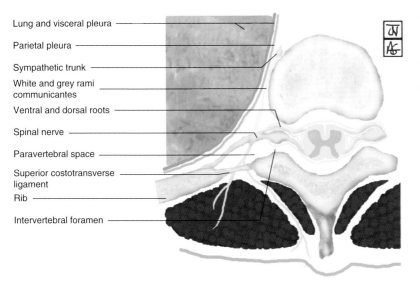

Figure 5.9 Transverse section of the paravertebral space at the level of the intervertebral foramen. The level of the section means that the head and neck of the rib are not seen. The superior costotransverse ligament passes between the neck of the rib and the transverse process superior to it, and may be appreciated as a 'click' as a needle passes into the paravertebral space.

Thoracic paravertebral block

Introduction

Paravertebral blockade may be performed at the cervical, thoracic or lumbar level but is most commonly performed at the thoracic level, because of the continuity of the paravertebral space at this level. A single high-volume low-concentration injection may provide unilateral ipsilateral blockade of 1–4 dermatomes. If anaesthesia of a greater area is required then multiple smaller-volume injections may be used.

Indications

Anaesthesia or analgesia for thoracic, cardiac and upper abdominal surgery, and for the acute pain of multiple fractured ribs. It has been associated with a decreased rate of malignant recurrence after breast surgery.[1]

Specific contraindications

- Intrathoracic infection.
- Empyema.
- Malignancy occupying the paravertebral space.
- Previous total pleurectomy.
- Kyphoscoliosis and previous surgery – increased risk of pleural puncture (altered anatomy, fibrosis or adhesions).

Pre-procedure checks

The patient is in a sitting, lateral or prone position. The most prominent cervical spinous process corresponds to C7, the spine of the scapula points to T2, and the lower border of the scapula corresponds to T7. Multiple injections are uncomfortable, and low-dose sedation may be required. The procedure is performed with an aseptic technique.

Technique

Needle: a paediatric 21G Tuohy is useful as it has 0.5 cm markings. An 18G Tuohy is an alternative.

Landmark

The superior aspect of the spinous process is identified and the location of the transverse process is marked, 2.5 cm laterally and 1 cm inferiorly. A paediatric Tuohy needle is inserted perpendicular to the skin and advanced to contact the transverse process at a maximum depth of 4 cm. If the process is not contacted then the needle is withdrawn and re-angled cranially or caudally. When the transverse process is contacted the needle is 'walked off' caudally below the process and advanced no more than 1 cm. At T8 and below the rib articulates more superiorly on the transverse process, so 'walking' cranially risks advancing onto the rib and then into the pleura.

Penetration of the superior costotransverse ligament (Figure 5.9) may be detected by loss of resistance,[2] but it is less well detected than that during an epidural placement, so it is not recommended to advance more than 1.5 cm. 15–30 ml of local anaesthetic is injected, after aspiration for blood, CSF, pleural effusion and air.

Ultrasound-assisted block

Several approaches have been described, none of which has yet been proven to be superior. The nerves themselves are not seen, but the boundaries of the paravertebral space containing the nerves are visualised and vascular structures may be identified with colour Doppler.

Parasagittal transducer position[3]

This approach is analogous to the landmark technique (above). A high-frequency probe is placed 5–10 cm lateral to the midline in a longitudinal plane overlying the ribs. The desired level is identified by visualising the first rib and counting down to the required level. The probe is moved medially, to a position 2.5 cm lateral to the midline, until the transverse process is identified by observing a change from semicircular-shaped ribs and well-visualised pleura to more superficial square-shaped transverse processes with less distinct pleura (Figure 5.10). The needle is inserted in-plane towards the inferior aspect of the transverse process and advanced into the space between the superior costotransverse ligament and the pleura. 20 ml of local anaesthetic is injected in aliquots after negative aspiration, and the pleura is seen to displace downwards.

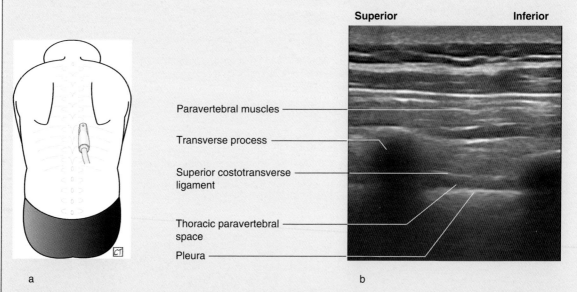

a   b

Figure 5.10 Parasagittal ultrasound image of the thoracic paravertebral space. The superior costotransverse ligament is illustrated in Figure 5.9.

Oblique subcostal transducer position[4]

The desired level and transverse processes are identified with a high-frequency probe, as described above. The probe is then placed in a transverse plane overlying the transverse process with the lateral end of the probe slightly rotated into an oblique position overlying and parallel to the rib. The transverse process is kept in view on the screen and the transducer is moved slightly caudally below the rib until an intercostal view is identified, indicated by visualisation of the external and internal intercostal muscles and the parietal pleura (Figure 5.11). The needle is advanced in-plane from lateral to medial through the external and internal intercostal membrane until the tip is positioned posterior (superficial) to the parietal pleura. A pop may be felt as the internal intercostal membrane is penetrated. 20 ml of local anaesthetic is injected in aliquots after negative aspiration, causing a downward displacement of the pleura.

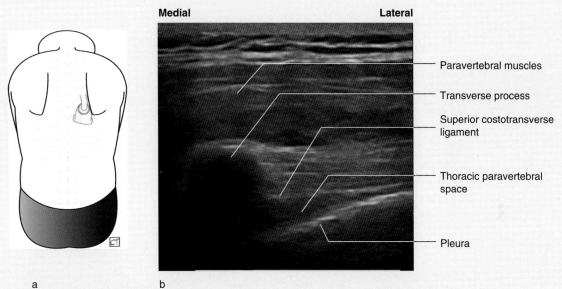

Medial   Lateral

Paravertebral muscles

Transverse process

Superior costotransverse ligament

Thoracic paravertebral space

Pleura

a   b

Figure 5.11 Oblique ultrasound image of the thoracic paravertebral space. The superior costotransverse ligament is illustrated in Figure 5.9.

Complications

Complications are relatively uncommon (2.6–5%):
- Hypotension.
- Pleural puncture and pneumothorax (onset may be delayed).
- Intravascular injection.
- Systemic local anaesthetic toxicity.
- Inadvertent epidural, subdural, intrathecal or spinal anaesthesia.
- Transient Horner's syndrome (T1–2).

Post-procedure checks

- The block should be assessed with pinprick or ice 10–15 minutes later and additional levels blocked as required.

References

1. Exadaktylos A, Buggy D, Moriarty D, *et al*. Can anesthetic technique for primary breast cancer affect recurrence or metastasis? *Anesthesiology* 2006; **105**: 660–4.
2. Eason M, Wyatt R. Paravertebral thoracic block: a reappraisal. *Anaesthesia* 1979; **34**: 638–42.
3. Luyet C, Eichenberger U, Greif R, *et al*. Ultrasound-guided paravertebral puncture and placement of catheters in human cadavers: an imaging study. *Br J Anaesth* 2009; **102**: 534–9.
4. Renes S, Bruhn J, Gielen M, *et al*. In-plane ultrasound-guided thoracic paravertebral block. *Reg Anesth Pain Med* 2010; **35**: 212–16.

The diaphragm

The diaphragm separates the thoracic and abdominal cavities and is the main muscle of respiration.

1. Embryology

Formed by fusion of:
- The septum transversum (mesoderm), which forms the central tendon.
- The dorsal oesophageal mesentery.
- The pleuro-peritoneal membranes.
- A peripheral rim of tissue originating from the body wall.

2. Structure

The curvature forms the right and left domes.
At end expiration, the right dome ascends to the fourth intercostal space, whereas the left dome reaches the fifth rib.
Composed of a central trefoil tendinous portion (blends with the fibrous pericardium) and a peripheral muscular portion with the following origins:
- Costal – inferior six costal cartilages.
- Xiphoid.
- Crura – left crus from bodies of L1 and 2, right crus from the bodies of L1, 2 and 3.

Fibres radiate towards central tendon.
There are three arcuate ligaments:
- Medial – a thickening of psoas fascia.
- Median – a fibrous arch between two crura.
- Lateral – a thickening of the lumbar fascia.

3. Diaphragmatic openings and structures passing within them

Refer to Figure 5.12.
T8 – inferior vena cava and right phrenic nerve. Opens in the central tendon, and thus respiration assists in venous return.
T10 – oesophagus, anterior and posterior vagal trunks, oesophageal branch of left gastric artery and vein.
T12 – aorta, azygous vein, thoracic duct. Found behind the median arcuate ligament (so opens behind the diaphragm not within it).
Mnemonic – COAL (Cava T8, Oesophagus T10, Aorta T12).
Lesser openings and structures passing within them:
- Between costal and xiphoid origins – superior epigastric vessels.
- Behind medial arcuate ligament – sympathetic trunks.
- Through the crura – greater, lesser and least splanchnic nerves.
- Through the left crus – hemiazygous vein.
- Adjacent to the inferior vena caval opening – right phrenic nerve.
- Adjacent to the pericardial attachment – left phrenic nerve.

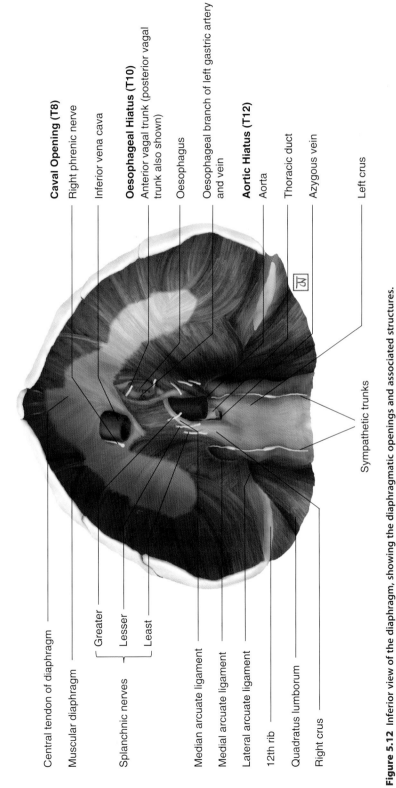

Central tendon of diaphragm

Muscular diaphragm

Greater
Lesser
Least
Splanchnic nerves

Median arcuate ligament

Medial arcuate ligament

Lateral arcuate ligament

12th rib

Quadratus lumborum

Right crus

Caval Opening (T8)

Right phrenic nerve

Inferior vena cava

Oesophageal Hiatus (T10)

Anterior vagal trunk (posterior vagal trunk also shown)

Oesophagus

Oesophageal branch of left gastric artery and vein

Aortic Hiatus (T12)

Aorta

Thoracic duct

Azygous vein

Left crus

Sympathetic trunks

Figure 5.12 Inferior view of the diaphragm, showing the diaphragmatic openings and associated structures.

4. Neurovascular supply

i) Arterial

The superior and inferior phrenic arteries (branches of the aorta).
Branches of the internal thoracic arteries.

ii) Venous

The superior and inferior phrenic veins (drain to the inferior vena cava or the azygous vein).
The internal thoracic veins.

iii) Nervous

The motor supply to each half of the diaphragm is from respective phrenic nerves ('C3, 4, 5 keeps the diaphragm alive').
The phrenic nerves form at the lateral border of the anterior scalene muscle, and descend on it with the internal jugular vein.
They enter the thorax posterior to the internal jugular vein at its union with the subclavian vein.
The right phrenic passes along the venous structures – superior vena cava, pericardium superficial to the right atrium, then inferior vena cava. The left phrenic passes over arterial structures –arch of the aorta, pericardium superficial to left atrium and left ventricle.
Both nerves pass anterior to the root of the lung.
Both phrenic nerves then pierce the diaphragm and innervate it from below.
The sensory supply to the central tendon is also from the phrenic nerves; to the muscular regions it is from the lower thoracic nerves.

5. Muscles of respiration

The primary muscles of respiration are the diaphragm and the intercostal muscles.
The following (accessory) muscles may contribute to respiration under forceful conditions:
- Pectoralis major and minor.
- Serratus anterior and posterior.
- The scalene muscles.
- Levatores costarum.

Thoracic contents: the trachea

Extends from C6 to T4 (15 cm long in the adult).
C-shaped cartilages are united by fibro-elastic tissue that is deficient posteriorly.
The trachealis muscle closes the posterior border and abuts the oesophagus.
Dives quite steeply posteriorly after passing through the thoracic inlet.

1. Relations

i) Anterior

Inferior thyroid veins, thymus remnants, brachiocephalic artery, left brachiocephalic vein, left common carotid artery, aortic arch.

ii) Posterior

Oesophagus.

iii) Right lateral

Right recurrent laryngeal nerve, pleura, azygous vein, right vagus nerve.

iv) Left lateral

Left recurrent laryngeal nerve, pleura, left common carotid artery, left vagus nerve.

2. Neurovascular supply

i) Arterial

Inferior thyroid artery (from the subclavian).

ii) Venous

Inferior thyroid veins (drain into the brachiocephalic veins).

iii) Nervous

Recurrent laryngeal nerves (derived from the vagus nerves) and the middle cervical ganglion (sympathetic).

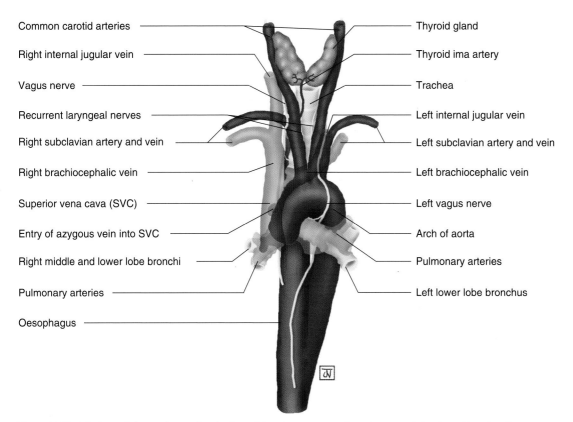

Common carotid arteries

Right internal jugular vein

Vagus nerve

Recurrent laryngeal nerves

Right subclavian artery and vein

Right brachiocephalic vein

Superior vena cava (SVC)

Entry of azygous vein into SVC

Right middle and lower lobe bronchi

Pulmonary arteries

Oesophagus

Thyroid gland

Thyroid ima artery

Trachea

Left internal jugular vein

Left subclavian artery and vein

Left brachiocephalic vein

Left vagus nerve

Arch of aorta

Pulmonary arteries

Left lower lobe bronchus

Figure 5.13 Relations of the trachea and major bronchi to the great vessels and nerves of the chest. Note that the recurrent laryngeal nerves depart the vagus nerve by looping around the subclavian artery on the right and around the arch of the aorta on the left. Both left and right recurrent laryngeal nerves lie in the tracheo-oesophageal groove.

Thoracic contents: the bronchial tree

1. Main bronchi

The trachea bifurcates at the T4 level, dividing into the left and right main bronchi.
The last tracheal ring is larger than the others, forming a bulge to the left of the midline known as the carina. The distance from the teeth to the carina is 20–25 cm in the adult.

- The right main bronchus is shorter, wider and more vertical than the left. It is 1.6–2.5 cm long and is angled at 25° to the vertical.
- The left main bronchus is 5.0–5.4 cm long and is angled at 45° to the vertical.

2. Relations

i) Right main bronchus

Superior – azygous vein.
Inferior then anterior – right pulmonary artery.

ii) Left main bronchus

Superior – aortic arch.
Posterior – thoracic duct, oesophagus, descending aorta.
Superior then anterior – left pulmonary artery.

3. Bronchial divisions

The right main bronchus divides to form the upper, middle and lower lobar bronchi.

The left main bronchus divides to form the upper and lower lobar bronchi.

Each lobar bronchus supplies a respective lung lobe. Each lobar bronchus further divides to form segmental bronchi, which supply the bronchopulmonary segments. There are 10 bronchopulmonary segments on the left and 10 on the right, as shown in Figure 5.14.

The superior bronchopulmonary segment of the lower lobe (of both lungs) is the most superior subdivision to arise from the posterior surface of the bronchial tree. Thus fluid in the bronchial tree acting by gravity alone may enter this bronchopulmonary segment in the supine patient.

Layers of the bronchial wall:

- Mucosa (contain ciliated and goblet cells, the latter reducing in number in the smaller bronchi).
- Basement membrane.
- Submucous layer (provides elastic recoil).
- Muscular layer (smooth muscle).
- Cartilage.

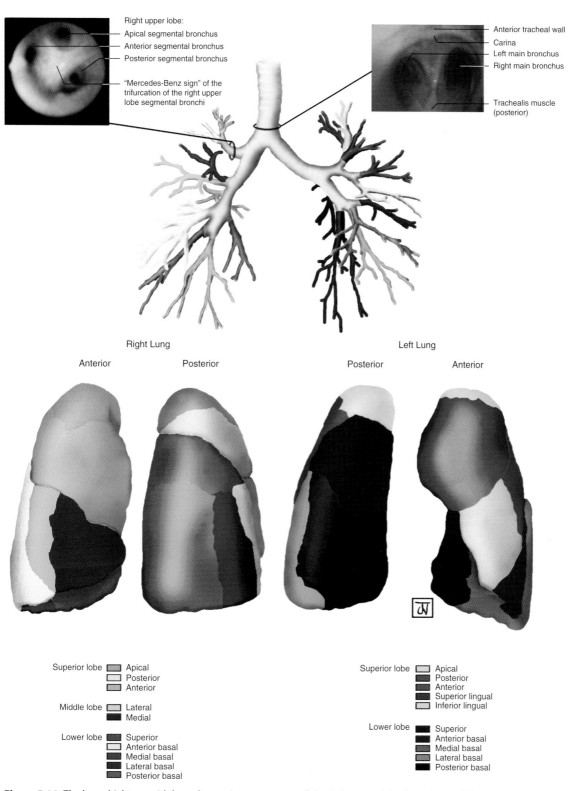

Right upper lobe:
- Apical segmental bronchus
- Anterior segmental bronchus
- Posterior segmental bronchus

"Mercedes-Benz sign" of the trifurcation of the right upper lobe segmental bronchi

- Anterior tracheal wall
- Carina
- Left main bronchus
- Right main bronchus

- Trachealis muscle (posterior)

Right Lung

Anterior   Posterior

Left Lung

Posterior   Anterior

Superior lobe
- Apical
- Posterior
- Anterior

Middle lobe
- Lateral
- Medial

Lower lobe
- Superior
- Anterior basal
- Medial basal
- Lateral basal
- Posterior basal

Superior lobe
- Apical
- Posterior
- Anterior
- Superior lingual
- Inferior lingual

Lower lobe
- Superior
- Anterior basal
- Medial basal
- Lateral basal
- Posterior basal

Figure 5.14 The bronchial tree, with bronchoscopic appearances of the right upper lobe bronchus and the carina.

4. Further subdivisions of the bronchial tree

Irregular branching of the respiratory tree results in 16 generations of conducting airways and a further 7 generations of gas-exchanging airways, giving a total of 23 airway generations.

The major bronchi account for generations 2–4 and are described above.

- Generations 5–11: Segmental bronchi, where cross-sectional area and therefore gas velocity fall.
- Generations 12–16: Bronchioles, where cartilage disappears and the proportion of smooth muscle rises. The epithelium changes from ciliated to cuboidal.
- Generations 17–19: Respiratory bronchioles, where cilia disappear and some alveoli bud off.
- Generations 20–23: Alveolar ducts and sacs. Alveoli number in the order of 300 million in a healthy adult.

5. Neurovascular supply

i) Arterial

Bronchial arteries (two on the left from the aorta, one on the right from the third right posterior intercostal artery).

ii) Venous

Bronchial veins (drain into the azygous on the right and the hemiazygous on the left) and the pulmonary veins (thus contributing to the fixed shunt of arterial blood).

iii) Nervous

As for the nerve supply to the lung, outlined in *The lungs*, section 3, below.

5. The pleura and mediastinum

i) The pleura

The lungs are enveloped by a double layer of pleura (imagine pushing the lung into the side of an inflated balloon – the pleura – such that the lung is completely enveloped by it):

- The visceral pleura, which is adherent to the lung and extends into the lung fissures.
- The parietal pleura, which lines the thoracic wall and the thoracic surface of the diaphragm.

The two pleural layers meet at the hilum forming the pulmonary ligament.

The potential space between the two pleural layers, the pleural cavity, contains a thin film of serous (pleural) fluid.

The presence of air between the visceral and parietal pleura constitutes a pneumothorax.

ii) The mediastinum

The space between the left and right pleural sacs is the mediastinum.

It is divided into superior and inferior portions by the sternal angle (of Louis).

The inferior mediastinum is further subdivided into anterior, middle and posterior portions by the pericardium.

Thoracostomy and chest drain insertion

Introduction

The pleural space may be drained by:

- A surgical opening made by blunt dissection then left open (thoracostomy). In the prehospital setting this may be all that is needed to stabilise a patient prior to transfer.
- A small-bore drain inserted via an image-guided Seldinger technique. This is the recommended first-line therapy, with a lower risk of complications and greater patient comfort.[1]
- Large-bore drains inserted by blunt dissection. Used where a small-bore drain fails or is likely to block, e.g. haemothorax or post-surgery.

Ultrasound is strongly recommended to determine the presence, location and size of pleural effusions and to look for loculation. The marking of a site with ultrasound and subsequent remote aspiration is not recommended.[1]

Pneumothorax may limit ultrasound examination, because of the poor transmission of sound waves through air.

Indications

The symptomatic treatment of pneumothorax, large pleural effusions, empyemas, chylothorax, traumatic haemothorax and post-surgery.

Specific contraindications

Relative:
- Uncorrected bleeding diathesis (INR > 1.5) unless tension pneumothorax.
- Chest wall infection at puncture site.

Pre-procedure checks

Recent chest x-ray to confirm side (with the exception of a tension pneumothorax).
Careful patient positioning.
Local anaesthetic with adrenaline is advised to reduce bleeding and increase patient comfort. Full aseptic technique.

Technique

Landmark/ultrasound-assisted technique

The patient is sat up, leaning forward, with arms on a pillow placed on an adjustable bedside table. Alternatively the patient is supine, positioned at the edge of the bed with the ipsilateral hand behind the head and a towel roll under the contralateral shoulder. This provides dependent drainage and access to the posterior axillary space.

Ultrasound is used to confirm the optimal puncture site, or the approach is made through the 'triangle of safety', an area bordered anteriorly by the lateral edge of pectoralis major, posteriorly by the lateral edge of latissimus dorsi, inferiorly by the line of the fifth intercostal space, and superiorly by the base of the axilla.

The skin is cleaned with 2% chlorhexidine gluconate in 70% isopropyl alcohol and covered with a sterile drape. The skin, subcutaneous tissue, rib periosteum, intercostal muscle and parietal pleura are infiltrated with local anaesthetic. If pleural fluid is not obtained via aspiration during anaesthetic infiltration then further imaging is required.

a

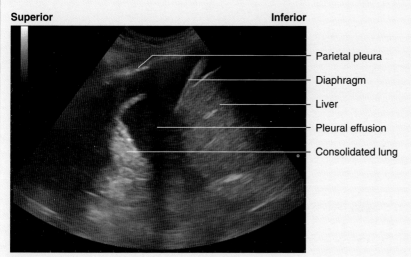

Superior · Inferior

- Parietal pleura
- Diaphragm
- Liver
- Pleural effusion
- Consolidated lung

b

Figure 5.15 Right pleural effusion with consolidated lung (curvilinear probe). Having identified the liver by its venous architecture, the diaphragm is noted superior to it. Only then can the lung and pleural space be assessed. Here an anechoic pleural effusion is noted with the lung floating within it. The lung itself is consolidated, as noted by its hyperechoic signal in this image.

Small-bore Seldinger technique

The needle is inserted over the superior aspect of the rib, avoiding the neurovascular bundle, within the triangle of safety. Following successful aspiration, the needle is stabilised and the guidewire is inserted into the pleural space. The needle is removed with the guidewire held in place. The dilator is threaded over the wire and the skin and tract dilated. The dilator is removed and the drain is inserted over the guidewire. Once the drain is in the pleural space, the guidewire is removed, the drain is sutured to the skin, a transparent sterile adhesive dressing is applied, and the drain is connected to the drainage system.

Large-bore blunt dissection

A 2–3 cm incision is made above and parallel to a rib within the triangle of safety. Blunt dissection is used to breach the intercostal muscles and parietal pleura before digital examination of the pleural space is undertaken to identify the lung and avoid other important structures. The large-bore drain is advanced over the superior aspect of the rib without the use of a trocar, avoiding the neurovascular bundle, until pleural fluid is obtained. It is secured with a non-purse-string suture and connected to a drainage system.

Complications

- Pneumothorax.
- Procedure failure – drain blockage.
- Pain and haemorrhage.
- Visceral injury (lung, liver, spleen, heart).
- Infection.
- Empyema.
- Damage to the intercostal and internal mammary vessels.
- Re-expansion pulmonary oedema – uncommon if a maximum of 1.5 litres of fluid is removed.
- Air embolism.

Post-procedure checks

- Chest x-ray to confirm drain position and exclude pneumothorax.
- CT chest if malposition suspected.

References

1. Havelock T, Teoh R, Laws D, *et al*. Pleural procedures and thoracic ultrasound: British Thoracic Society pleural disease guideline 2010. *Thorax* 2010; **65** (Suppl 2): 61–76.

Thoracic contents: the lungs

The lung lobes and bronchopulmonary segments are outlined under *The bronchial tree*, above.

1. Fissures

The lobes of the lung are separated by fissures:
- Oblique fissure – present in the right and left lungs. Separates middle and lower lobes (from T2 spinous process to the sixth costochondral junction).
- Horizontal fissure – present in the right lung only. Separates upper and middle lobes (from the fourth costochondral junction to the oblique fissure).

2. Lung aspects

Each lung has an:
- Apex – extends 2.5 cm above the midpoint of the clavicle, so is at risk during needling of the neck or trauma above the clavicle.
- Base – concave, but more so on the right (due to the liver).
- Hilum – contains structures entering and leaving the lung. On each side: one bronchus, one pulmonary artery, two pulmonary veins, bronchial vessels, nerve plexuses and lymph nodes. The right upper lobe bronchus (with its accompanying pulmonary artery) leaves the right main bronchus proximal to the hilum.

Each lung has a costal, mediastinal and diaphragmatic surface.

The cardiac notch is a defining feature of the left lung. The inferior borders of the lung, midway between quiet inspiration and expiration, are defined by:

- The sixth rib in the mid-clavicular line.
- The eighth rib in the mid-axillary line.
- The tenth rib in the mid-scapular line.

  At extremes of respiration, lung excursion may vary by 5–8 cm.

3. Neurovascular supply

The bronchial circulation, which supplies the stroma of the lung, bronchi and lymph nodes, is described under *The bronchial tree*, section 5, above.

i) Arterial

Pulmonary artery (containing deoxygenated blood).

ii) Venous

Pulmonary veins (two from each lung, containing oxygenated blood).

iii) Nervous

Pulmonary plexus (derived from the cardiac plexus). Parasympathetic fibres are afferent (cough reflex – used in brainstem death testing) and efferent (smooth muscle constriction of bronchi and pulmonary arterioles). Sympathetic fibres are dilator to pulmonary arterioles and bronchi.

iv) Lymph

Drains to bronchopulmonary then tracheobronchial nodes, and from there to either the thoracic duct (on the left) or the lymphatic duct (on the right).

Point-of-care lung ultrasound

Lung ultrasound is an invaluable point-of-care tool in the rapid assessment of critically unwell patients. It can most usefully be applied in the rapid inclusion or exclusion of pneumothorax, interstitial syndromes (such as pulmonary oedema), lung consolidation and pleural effusion (Figure 5.15). It may also be used as a tool to sequentially monitor the effects of therapy on the lung.

Pneumothorax is arguably one of the simplest and most useful diagnoses to make with lung ultrasound. It is more sensitive and more specific then chest x-ray and is the investigation of choice in an unstable patient. With the supine patient, examination in the mid-clavicular line is likely to be most informative. Diagnostic features of pneumothorax on lung ultrasound are:

1) Absence of lung sliding (regular movement between visceral and parietal pleura – 'pearls on a string').
2) Absence of B-lines (vertical hyperechoic lines ('comet tails') arising from the visceral pleura, which extend to the bottom of the screen without fading and move with lung sliding). Interposition of air between parietal and visceral layers prevents B-line formation.
3) Absence of lung pulse (rhythmic movement of the visceral upon the parietal pleura with cardiac oscillations).
4) Presence of the lung point(s). Defines the limit of a pneumothorax and is the point of transition between the presence and absence of lung sliding and B-lines.

The presence of points 1–3 is sufficient to diagnose pneumothorax in an emergency.

M-mode (to identify the 'seashore sign' – a clear distinction between tissue superficial and deep to the pleura) may be used to support the above findings.

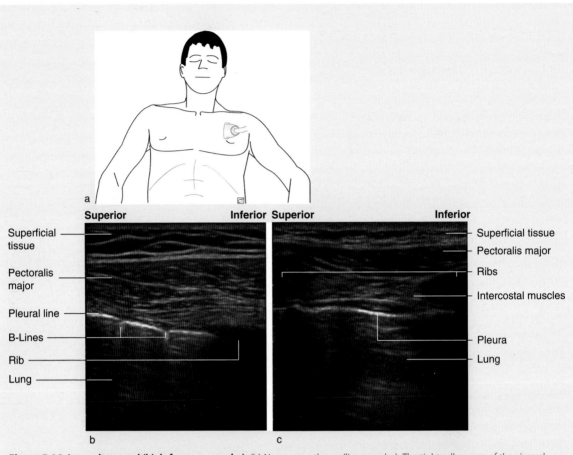

Figure 5.16 Lung ultrasound (high-frequency probe). (b) No pneumothorax (linear probe). The tight adherence of the visceral and parietal layers of pleura generates B-lines ('comet tails') which move throughout the respiratory cycle ('pearls on a string').
(c) Pneumothorax (linear probe). Image from the contralateral side of the same patient. The presence of air between the visceral and parietal layers of pleura prevents B-line formation. Although air may not be visualised, the absence of B-lines is a key finding when noted in the presence of the other findings described in the text.

References

1. Volpicelli G, Elbarbary M, Blaivas M, *et al*. International evidence-based recommendations for point-of-care lung ultrasound. *Int Care Med* 2012; **38**: 577–91.
2. Lichtenstein DA, Mezière GA. Relevance of lung ultrasound in the diagnosis of acute respiratory failure: the BLUE protocol. *Chest* 2008; **134**: 117–25.

Thoracic contents: the heart

Four-chamber conical muscular pump in the middle mediastinum.

1. Pericardium

Surrounds the heart.
Divided into:

i) Fibrous pericardium

Encloses the heart and roots of the great vessels.
Blends with central tendon of diaphragm.
Prevents over-distension of the heart.
Supplied by the phrenic nerve.

ii) Serous pericardium

Divided into parietal and visceral layers, with a potential space between the two.
The parietal layer lines the inner surface of the fibrous pericardium and is supplied by the phrenic nerve.
The visceral layer is in direct contact with the heart and is insensitive.

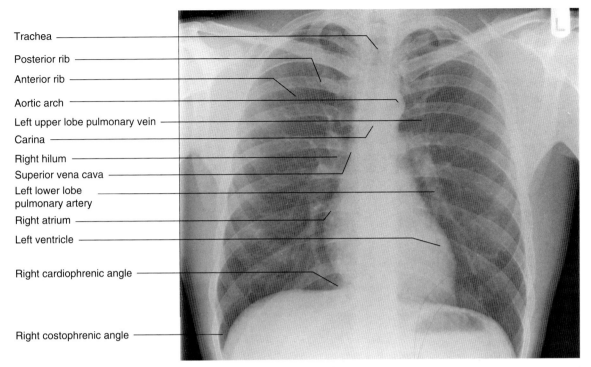

Trachea
Posterior rib
Anterior rib
Aortic arch
Left upper lobe pulmonary vein
Carina
Right hilum
Superior vena cava
Left lower lobe pulmonary artery
Right atrium
Left ventricle
Right cardiophrenic angle
Right costophrenic angle

Figure 5.17 The important features of a normal posteroanterior (PA) chest x-ray.

2. Borders

Right – right atrium.
Left – left auricle, left ventricle.
Anterior – right ventricle.
Posterior – left atrium, some right atrium.
Diaphragmatic – right and left ventricles.

3. Chambers

i) Right atrium

Receives blood from the superior and inferior venae cavae and the coronary sinus.
Pumps blood to the right ventricle via the tricuspid valve.

ii) Right ventricle

Pumps blood ejected by the right atrium into the pulmonary trunk via the pulmonary valve, which has three semilunar cusps.
Internally muscular ridges (trabeculae carneae) are seen, which condense to form three papillary muscles. These muscles attach to the ventricular surfaces of the tricuspid valve via chordae tendinae and prevent inversion of the valve during systole.

iii) Left atrium

Receives oxygenated blood from the lungs via four valveless pulmonary veins.
Ejects blood via the bicuspid mitral valve into the left ventricle.
Its wall is thicker than that of the right atrium.

iv) Left ventricle

Thick-walled chamber (twice as thick as the right ventricle) responsible for ejection of blood via the aortic valve, which has three semilunar cusps.
Trabeculae carneae again condense to form two papillary muscles, which attach via chordae tendinae to the cusps of the mitral valve.
Superior to the aortic valve cusps lie the aortic sinuses, which are dilations of the aortic wall. The left and right aortic sinuses give rise to respective coronary arteries; the posterior aortic sinus does not give rise to a coronary artery and hence is known as the non-coronary sinus.

Anterior View

Posterior View

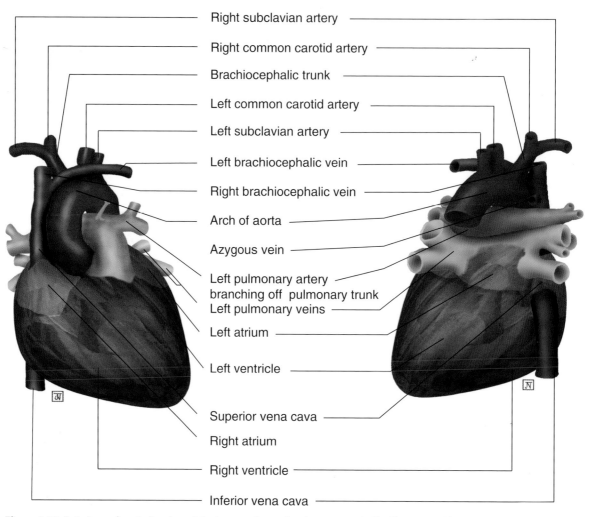

Right subclavian artery

Right common carotid artery

Brachiocephalic trunk

Left common carotid artery

Left subclavian artery

Left brachiocephalic vein

Right brachiocephalic vein

Arch of aorta

Azygous vein

Left pulmonary artery
branching off pulmonary trunk

Left pulmonary veins

Left atrium

Left ventricle

Superior vena cava

Right atrium

Right ventricle

Inferior vena cava

Figure 5.18 Anterior and posterior view of the heart and associated great vessels. The fibrous strand between the aorta and the pulmonary artery is the ligamentum arteriosum - the remnant of the fetal ductus arteriosus.

Focused echocardiogram

Focused echo is used to provide basic information on the size and function of the ventricles, basic valvular function, the presence of pericardial fluid and the filling status of the patient. It is also used sequentially to reassess the efficacy of therapy, particularly filling status. It is an invaluable bedside tool in the patient in shock or in cardiac arrest. Focused echo uses four views of the heart, using a phased array probe, and appreciating the anatomy helps the user to understand the images obtained. The views are:

(1) Parasternal long-axis.
(2) Parasternal short-axis.
(3) Apical four-chamber.
(4) Subcostal.

1. Parasternal long-axis view

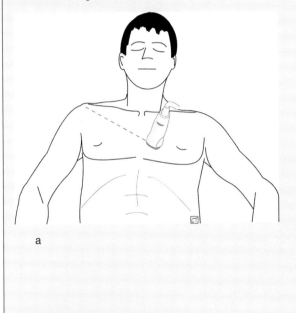

a

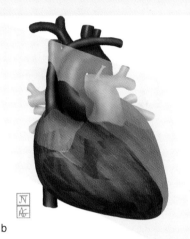

b

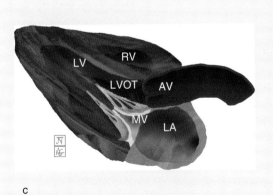

c

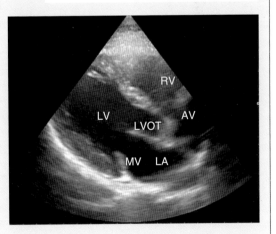

d

Figure 5.19 Parasternal long-axis view of the heart (phased array probe). (a) The probe is placed at the left sternal edge with the marker pointing to the right shoulder. The third or fourth intercostal space is used to visualise the heart in long axis. Turning the patient onto the left side may help obtain better views. (b) The beam of the ultrasound is shown by the 'pane' on the anatomical figure. (c, d) Long-axis view of the heart with associated ultrasound image. AV, aortic valve; LA, left atrium; LV, left ventricle; LVOT, left ventricular outflow tract; MV, mitral valve; RV, right ventricle.

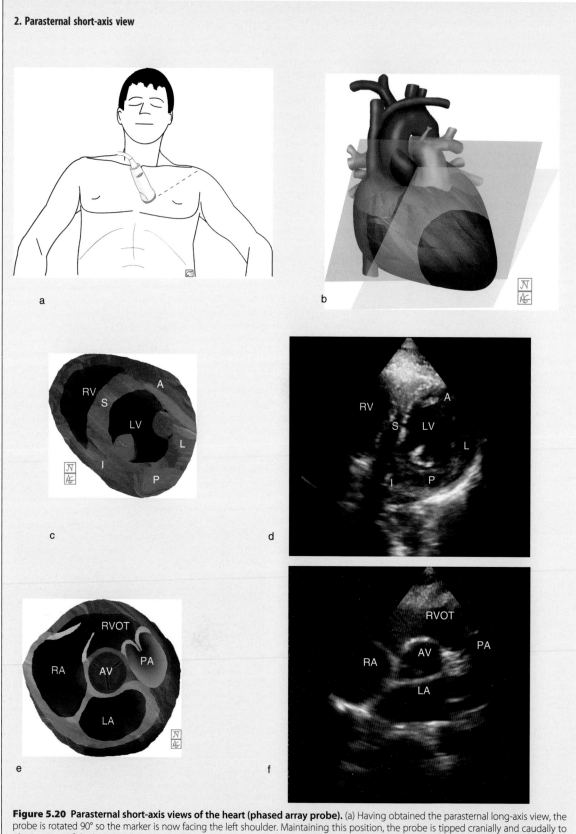

Figure 5.20 Parasternal short-axis views of the heart (phased array probe). (a) Having obtained the parasternal long-axis view, the probe is rotated 90° so the marker is now facing the left shoulder. Maintaining this position, the probe is tipped cranially and caudally to obtain views of the heart in short axis. (b) The beams of the ultrasound are shown by the 'panes' on the anatomical figure. (c-f) Short-axis views of the heart, correlating with the planes shown in (b), with associated ultrasound images. AV, aortic valve; LA, left atrium; LV, left ventricle; PA, pulmonary artery and pulmonary valve; RA, right atrium; RV, right ventricle; RVOT, right ventricular outflow tract; S, septum. Walls of the left ventricle: A, anterior; I, inferior; L, lateral; P, posterior.

3. Apical four-chamber view

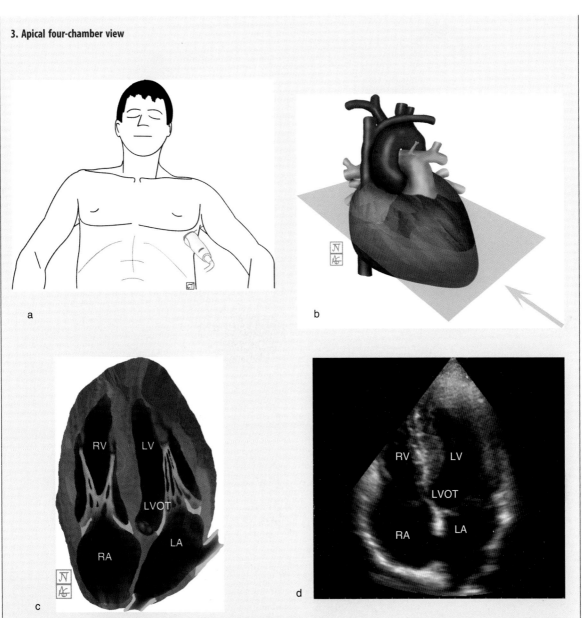

Figure 5.21 Apical four-chamber view of the heart (phased array probe). (a) The probe is placed at the apex beat with the marker facing the axilla. (b) The beam of the ultrasound is shown by the 'pane' on the anatomical figure. (c, d) Four-chamber view of the heart with associated ultrasound image. As the atrioventricular valves are in plane, this is a good view to look for valvular incompetence using Doppler. LA, left atrium; LV, left ventricle; LVOT, left ventricular outflow tract; RA, right atrium; RV, right ventricle.

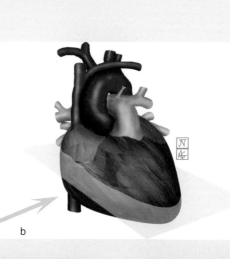

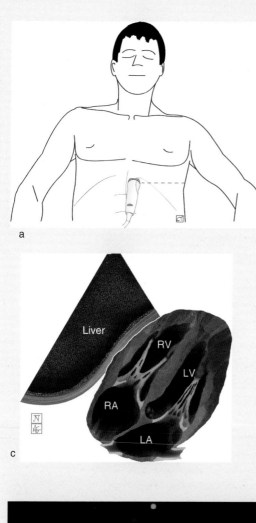

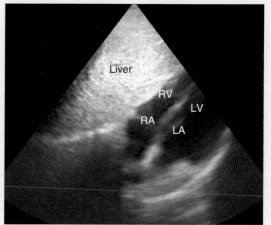

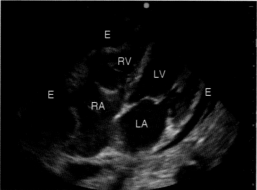

Figure 5.22 Subcostal views of the heart (phased array probe).
(a) The probe is placed beneath the xiphisternum with the marker pointing towards the patient's left (three o'clock). (b) The beam of the ultrasound is shown by the 'pane' on the anatomical figure. (c, d) Subcostal view of the heart with associated ultrasound image. (e) A pericardial effusion or tamponade may be most easily seen with a subcostal view, as shown here. The patient was known to have malignancy and was referred to critical care with unstable blood pressure. A large pericardial effusion is noted, with diastolic collapse of the right ventricle, confirming the diagnosis of pericardial tamponade. Emergency drainage was undertaken to avert cardiac arrest. LA, left atrium; LV, left ventricle; RA, right atrium; RV, right ventricle; E, pericardial effusion.

References

1. Breitkreutz R, Walcher F, Seeger FH. Focused echocardiographic evaluation in resuscitation management: concept of an advanced life support–conformed algorithm. *Crit Care Med* 2007; **35** (suppl): S150–61.
2. Bernard P. Cholley Vieillard-Baron A, Mebazaa A. Echocardiography in the ICU: time for widespread use. *Int Care Med* 2005; **32**: 9–10.
3. Orme RMLE, Oram MP, McKinstry CE. Impact of echocardiography on patient management in the intensive care unit: an audit of district general hospital practice. *Br J Anaesth* 2009; **102**: 340–4.

4. Conducting system

i) Sinoatrial (SA) node

Located in the right atrial wall near the superior vena cava opening. Depolarisation spreads across the atria, but is prevented from spreading to the ventricles by the fibrous atrioventricular ring.

ii) Atrioventricular (AV) node

At the base of the right atrial septal wall, near the tricuspid valve and the coronary sinus. The only route by which depolarisation may spread to the ventricles in health.

iii) Bundle of His

Nerve fibre bundle running from the atrioventricular node, within the interventricular septum, to the left and right ventricles.
The bundle terminates by dividing into left and right bundle branches, each of which supplies its respective ventricle.
The left bundle branch divides into anterior and posterior fascicles.

5. Neurovascular supply

Arterial supply is from two coronary arteries which arise from the left and right aortic sinuses, just superior to the aortic valve in the ascending aorta.

i) Right coronary artery

Arises from the right aortic sinus and descends between the pulmonary trunk and the right

atrium to run in the anterior atrioventricular groove.
Continues in the posterior atrioventricular groove to anastomose with the circumflex artery.
Branches:
- SA nodal (in 60%) and AV nodal (in 80%).
- Right marginal.
- Posterior interventricular.
Supplies:
- SA node (in 60%) and AV node (in 80%).
- Right atrium and ventricle.
- Part of the left ventricle.
- Posterior interventricular septum.

ii) Left coronary artery

Arises from the left aortic sinus and passes behind the pulmonary trunk.
Branches:
- Circumflex – gives off the left marginal branch and the SA nodal branch (in 40%).
- Anterior interventricular (also known as left anterior descending) – gives off the diagonal and obtuse marginal branches.
Supplies:
- SA node (in 40%) and AV node (in 20%).
- Left atrium and ventricle.
- Part of the right ventricle.
- Anterior interventricular septum.

The above describes the more common right-dominant circulation. In 10% of people the posterior interventricular artery is supplied by a continuation of the anterior interventricular artery, and this is described as a left-dominant circulation.

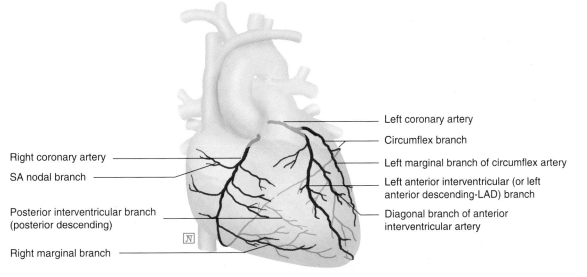

Figure 5.23 The arterial supply of the heart.

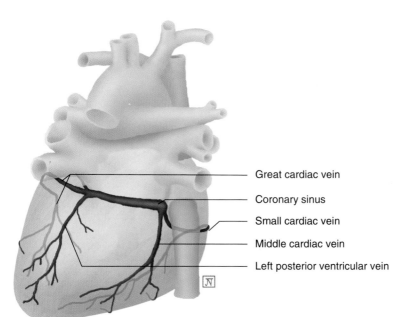

Figure 5.24 Posterior view showing the venous drainage of the heart.

Great cardiac vein

Coronary sinus

Small cardiac vein

Middle cardiac vein

Left posterior ventricular vein

iii) Venous

The coronary sinus carries two-thirds of the venous return and empties into the right atrium. It is fed by veins which accompany the arteries:

- The great cardiac vein – accompanies the anterior interventricular artery (hence drains left coronary territory).
- The middle cardiac vein – accompanies the posterior interventricular artery.
- The small cardiac vein – accompanies the marginal branch of the right coronary artery.

The middle and small cardiac veins drain right coronary territory.

The venae cordis minimae or Thebesian veins carry one-third of the venous return and drain directly into the four chambers of the heart (but mostly into the right atrium). These veins therefore contribute to the shunt fraction of arterial blood.

iv) Nervous

Supplied by the superficial and deep cardiac plexuses (mixed autonomic supply):

- Sympathetic – presynaptic fibre cell bodies are found in the lateral horns of the superior five or six thoracic segments of the spinal cord. Postsynaptic cell bodies are found in the three cervical ganglia and the superior four thoracic ganglia of the sympathetic trunk (see Chapter 1, section 6).
- Parasympathetic – from the left and right vagus nerves.

Thoracic contents: the great vessels

Refer to Figure 5.13.

1. Aorta

Commences at the aortic valve and terminates at L4 by bifurcation into the common iliac arteries.
Four parts:

i) Ascending

5 cm long. It gives off:
- Right and left coronary arteries.

ii) Arch

It gives off:
- Brachiocephalic trunk – divides into right subclavian and right common carotid arteries.
- Left common carotid artery.
- Left subclavian artery.
- (Occasionally the thyroid ima artery).

iii) Descending

Commences at T4 and terminates at the aortic opening in the diaphragm (T12). It gives off:
- Nine posterior intercostal arteries and one subcostal artery (at the twelfth rib).
- Bronchial and oesophageal arteries.

iv) Abdominal

Commences at the aortic opening of the diaphragm and ends by bifurcating into the

common iliac arteries at L4. It gives off (from superior to inferior):
- Coeliac trunk – supplies the foregut.
- Superior mesenteric artery – supplies the midgut.
- Left and right renal arteries – supply the kidneys.
- Left and right gonadal arteries – supply the testicle or ovary/ fallopian tube.
- Inferior mesenteric artery – supplies the hindgut.

Note: the foregut (and the organs associated with it) is found proximal to the second part of the duodenum, the hindgut is distal to a point two-thirds of the way along the transverse colon, and the midgut lies between the two.

2. Brachiocephalic trunk

Has no branches.
Divides into right subclavian and right common carotid arteries.

3. Common carotid arteries

Left and right. They have no branches.
Ascend in the carotid sheath and divide in the neck into the external and internal carotid arteries. These are described in more detail in Chapter 4, *The vessels of the neck*.

4. Subclavian arteries

Left and right. They pass over the apex of the lung, then over the first rib, posterior to the anterior scalene muscle. They become the axillary arteries at the lateral margin of the first rib.

5. Pulmonary arteries

The left pulmonary artery is attached to the underside of the aortic arch by the ligamentum arteriosum (the remnant of the ductus arteriosus – see Figure 5.18). It is found superior then anterior to the left main bronchus. The right pulmonary artery passes below the carina. It is found inferior then anterior to the right main bronchus. It gives off a branch to the upper lobe before entering the hilum.

6. Brachiocephalic veins

The left and right brachiocephalic veins are formed behind their respective sternoclavicular joints by the union of the internal jugular and subclavian veins.
- The right brachiocephalic vein is 3 cm long and passes vertically inferiorly.
- The left brachiocephalic vein is 6 cm long and passes almost horizontally behind the manubrium. It may project slightly above the jugular notch if the vein is distended, particularly if the head is extended, where it may be vulnerable during tracheostomy. It receives the thoracic duct.

7. Superior vena cava

Formed by the union of the left and right brachiocephalic veins at the lower border of the first right costal cartilage. It passes vertically inferiorly behind the right border of the sternum, where it receives the azygous vein before entering the right atrium.

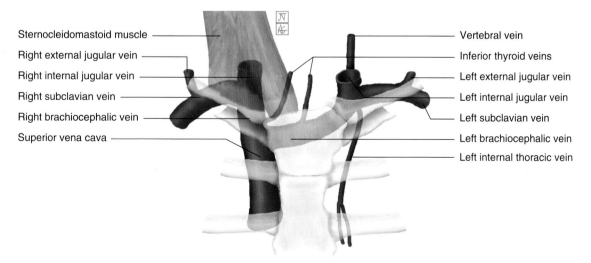

Figure 5.25 Large veins of the thorax/root of neck. Labels refer to the underlying venous structures only. Note how the internal jugular vein is found deep to the space between the two heads of sternocleidomastoid – a useful landmark for central venous access when performing a low approach to the vessel. Note also the proximity of the left brachiocephalic vein to the superior edge of the manubrium, placing it at risk of puncture (early bleed) or erosion (late bleed) from a low-lying tracheostomy.

The upper limb

The neurovascular supply of the upper limb is derived from the brachial plexus and the subclavian vessels, both of which pass over the first rib and beneath the clavicle to reach the limb.

1. Vascular supply of the upper limb

i) Arterial

See Figure 6.1.

Subclavian artery

Lateral border of first rib

Axillary artery

Inferior border of teres major muscle

Brachial artery

Radial artery

Ulnar artery

Deep palmar arch

Superficial palmar arch

Digital arteries

Figure 6.1 Schematic illustration of the major arteries of the upper limb. Labels in italics represent the transition points from one named artery to the next. The brachial artery divides into the radial and ulnar arteries in the cubital fossa.

ii) Venous

Variable.

Three main veins drain the upper limb (Figure 6.2):

- Basilic (medial).
- Cephalic (lateral).
- Median vein of the forearm.

The cephalic vein enters the deltopectoral triangle to join the axillary vein.

The basilic vein becomes the axillary vein, which becomes the subclavian vein. The transition points are the same as in the arterial tree.

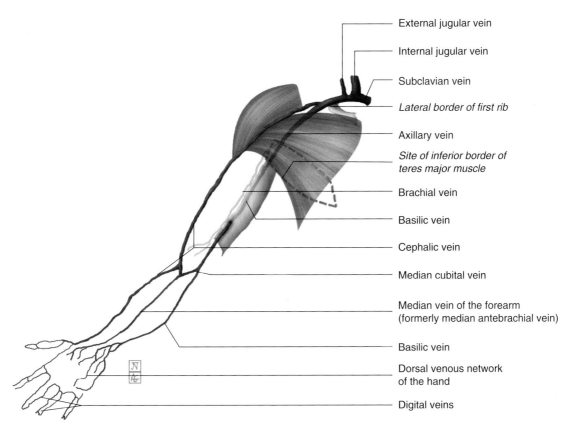

External jugular vein

Internal jugular vein

Subclavian vein

Lateral border of first rib

Axillary vein

Site of inferior border of teres major muscle

Brachial vein

Basilic vein

Cephalic vein

Median cubital vein

Median vein of the forearm (formerly median antebrachial vein)

Basilic vein

Dorsal venous network of the hand

Digital veins

Figure 6.2 Schematic illustration of the major veins of the upper limb. Labels in italics represent the transition points from one named vein to the next. Note that the cephalic vein begins in the anatomical snuffbox and runs the length of the upper limb, passing between the deltoid and pectoralis major muscles to drain into the axillary vein. The basilic vein begins at the medial aspect of the dorsal venous network and ascends on the medial aspect of the upper limb, passing through the deep fascia to become the axillary vein at the inferior border of the teres major muscle. It is joined shortly afterwards by the brachial vein.

2. The axilla

i) Definition

The axilla is a pyramidal-shaped area facilitating the passage of structures from the upper limb to the neck or thorax.

Its shape varies depending on the position of the arm; it almost disappears with the arm fully abducted.

It has the following borders:

- Apex – bounded by the clavicle, first rib and scapula/subscapularis muscle.
- Base – skin and axillary fascia between the arm and the thoracic wall.
- Anterior – pectoralis major and minor muscles.
- Posterior – scapula and subscapularis, teres major, latissimus dorsi muscles.
- Medial – upper four ribs and serratus anterior.
- Lateral – intertubercular groove of humerus.

ii) Contents

- Axillary artery.
- Axillary vein.
- Brachial plexus and terminal nerves (the cords of the brachial plexus are named with respect to their position about the axillary artery).
- Axillary lymph nodes – drain the lateral breast, chest wall and upper limb. Drain into the thoracic duct on the left and the lymphatic duct on the right.

Axillary brachial plexus block

Introduction

Axillary brachial plexus block is the technique of choice for distal upper limb procedures. A multi-nerve injection technique has a higher success rate and a shorter onset time than a single high-volume injection.[1–4]

Indications

Surgery to the forearm, wrist or hand of medium to long duration with or without a tourniquet.

Technique

Landmark

Transarterial technique

The arm is abducted to 90° and the elbow flexed with the forearm resting on a pillow (Figure 6.4a). The axillary artery is punctured with a 25G needle at the level of the anterior axillary fold, and the needle is advanced deeper until aspiration of arterial blood ceases. 20 ml of local anaesthetic is injected to block the radial nerve. The needle is withdrawn through the artery until aspiration of arterial blood ceases and 20 ml of local anaesthetic is injected to block the median and ulnar nerves. In theory, block of the musculocutaneous nerve occurs by spread of the local anaesthetic within the sheath, but in practice this is almost always missed. The radial nerve may also be missed with this technique.

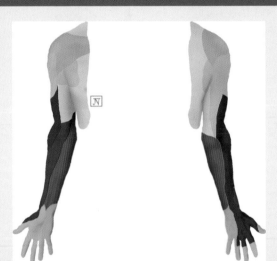

Figure 6.3 The distribution of anaesthesia following axillary brachial plexus block. The area of anaesthesia is shown by the coloured areas, which correlate with Figure 6.5.

Nerve stimulator

The arm is positioned as above and the artery palpated. The nerve stimulator is set to 0.5–1 mA (2 Hz). The following multi-injection technique carries the risk of inadvertent nerve damage, as spread of local anaesthetic from one injection location may partially block other nerves, leading to absent motor twitches.

Median nerve

The needle is inserted above the axillary artery, at 45° to the skin, aiming proximally, until a 'click' is felt as the brachial superficial fascia is penetrated. 5–10 ml of local anaesthetic is injected following the motor response of flexion of the first three digits.

Musculocutaneous nerve

From the above insertion point the needle is redirected subcutaneously into and above the coracobrachialis muscle. The motor response for the musculocutaneous nerve is elbow flexion. 5–10 ml of local anaesthetic is injected.

Ulnar nerve

The stimulating needle is inserted below the axillary artery, at 45° to the skin, aiming proximally until the brachial superficial fascia is penetrated. 5–10 ml of local anaesthetic is injected following the motor response of thumb adduction and flexion of the fifth digit.

Radial nerve

From the position described for the ulnar nerve block the needle is advanced deeper and slightly upward behind the artery until thumb extension is obtained. 5–10 ml of local anaesthetic is injected.

Spread of the local anaesthetic following radial nerve block may produce anaesthesia of the ulnar nerve without the need for a separate ulnar injection.[5]

Ultrasound-assisted block

The arm is in the position described above, and a high-frequency probe is placed in the axilla as shown in Figure 6.4a. The axillary artery and humerus are identified. The confluence of the tendons of latissimus dorsi and teres major is identified as a hyperechoic line seen inserting into the humerus. This conjoined tendon signifies the level at which to perform the block, promoting reliable spread of local anaesthetic around the nerves, particularly the radial nerve.[6–7] It is possible to alter the positions of the nerves relative to the axillary artery by flexing and extending the elbow. Figure 6.4b illustrates the most common locations of the nerves. The user is encouraged to confirm the identity of each nerve by tracing them from a more distal location in the arm (see box on *Wrist block*) up into the axilla. It is usually possible to block the median, ulnar and radial nerves with a single needle insertion point. The musculocutaneous nerve may be easier to visualise a few centimetres more distally in the arm, requiring a second needle insertion. The radial nerve is not always seen clearly at this level, because of the ultrasound artefact of posterior enhancement behind the artery.

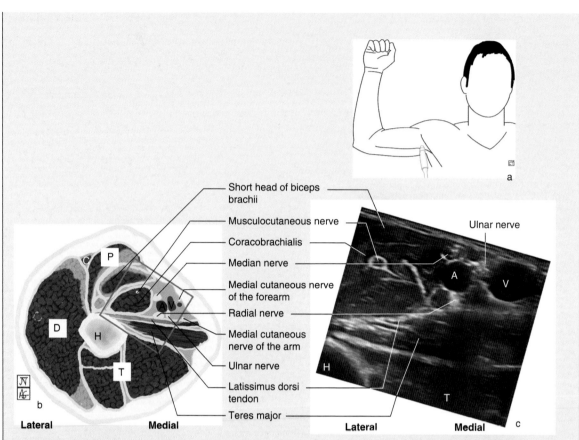

Figure 6.4 The anatomy of an axillary brachial plexus block and associated ultrasound image. Cross-sectional anatomy of the axilla at the level of the insertion of latissimus dorsi and teres major. The blue box depicts the area shown in the ultrasound image. C, cephalic vein; P, pectoralis major; D, deltoid; T, triceps; H, humerus; A, axillary artery; V, axillary vein. Note that in this particular ultrasound image the ulnar nerve is between the axillary vessels, which is a normal variant of the more common arrangement shown in the anatomical figure.

Complications

- Vascular puncture, intravascular injection or systemic absorption of local anaesthetic.
- Haematoma.
- Nerve injury.

References

1. De Jong R. Axillary block of the brachial plexus. *Anesthesiology* 1961; **22**: 215–25.
2. Vester-Andersen T, Broby-Johansen U, Bro-Rasmussen F. Perivascular axillary block VI: the distribution of gelatine solutions injected into the axillary neurovascular sheath of cadavers. *Acta Anaesthesiol Scand* 1986; **30**: 18–22.
3. Klaastad O, Smedby O, Thompson G, *et al*. Distribution of local anesthetic in axillary brachial plexus block: a clinical and magnetic resonance imaging study. *Anesthesiology* 2002; **96**: 1315–24.
4. Chin KJ, Handoll H. Single, double or multiple-injection techniques for axillary brachial plexus block for hand, wrist or forearm surgery in adults. *Cochrane Database Syst Rev* 2011; (**7**): CD003842.
5. Sia S, Bartoli M. Selective ulnar nerve localization is not essential for axillary brachial plexus block using a multiple nerve stimulation technique. *Reg Anesth Pain Med* 2001; **26**: 12–16.
6. Gray A. The conjoint tendon of the latissimus dorsi and teres major: an important landmark for ultrasound-guided axillary block [letter]. *Reg Anesth Pain Med* 2009; **34**: 179.
7. Buijze G, Keereweer S, Jennings G *et el*. Musculotendinous transfer as a treatment option for irreparable postero-superior rotator cuff tears: teres major or latissimus dorsi? *Clin Anat* 2007; **20**: 919–23.

3. Neural supply of the upper limb

The brachial plexus has been described in Chapter 4, *The nerves of the neck* section 2. The following more explicitly delineates the anatomy of the terminal nerves distal to the shoulder (Figure 6.8).

i) Radial nerve (C5–T1)

Continuation of the posterior cord.
Runs posterior to the axillary and brachial arteries before passing around the spiral groove of the humerus.
Enters the anterior compartment of the arm, and 2–3 cm proximal to the elbow divides into deep (mostly muscular) and superficial (entirely cutaneous) branches.
The superficial branch descends in the flexor compartment under brachioradialis before passing over the anatomical snuffbox to enter the hand (Figure 6.8).
The deep branch winds round the neck of the radius to descend in the posterior compartment, where it is known as the posterior interosseous nerve.
- Motor – extensors of the elbow, wrist and digits. Abducts the thumb.
- Sensory – superficial branch: skin of the posterior aspects of the upper limb (posterior cutaneous nerves of the arm and forearm), skin over the snuffbox and lateral posterior hand. Deep branch (posterior interosseous nerve): periosteum of the radius, ulnar and carpal joints.

ii) Axillary nerve (C5–6)

Continuation of the posterior cord.
Found posterior to the axillary artery.
Wraps around the neck of the humerus.
- Motor – deltoid and teres minor.
- Sensory – skin over inferior part of deltoid.

iii) Musculocutaneous nerve (C5–7)

Continuation of the lateral cord.
Runs lateral to the axillary artery then pierces coracobrachialis before descending between biceps and brachialis muscles.

Terminates as the lateral cutaneous nerve of the forearm.
- Motor – flexors of the elbow joint.
- Sensory – skin over the lateral forearm.

iv) Median nerve (C6–T1)

Arises from the medial and lateral cords.
Anterior then lateral to the axillary artery.
Crosses the brachial artery at mid-humeral level to lie medial to it in the cubital fossa. Shortly afterwards, gives off the anterior interosseous nerve, which runs with the corresponding artery on the interosseous membrane to supply muscles of the forearm and wrist joint.
Lies between flexor carpi radialis and palmaris longus before passing through the carpal tunnel.
Gives off a cutaneous branch prior to passing under the retinaculum and a recurrent branch after doing so.
Splits into lateral and medial branches before entering the hand.
Continues as digital nerves to the digits.
- Motor – in the wrist, all except flexor carpi ulnaris and half of flexor digitorum profundus. In the hand, the thenar muscles and lumbricals 1 and 2 (mnemonic – LOAF: Lumbricals 1 and 2, Opponens pollicis, Abductor pollicis brevis, Flexor pollicis brevis).
- Sensory – wrist joint. Skin of the lateral 3.5 digits and palm anteriorly and the distal 3.5 digits posteriorly (digital nerves).

v) Ulnar nerve (C8–T1)

Arises from the medial cord.
Runs on the medial side of the brachial artery.
Passes behind the medial epicondyle of the humerus then enters the forearm.
Runs deep to flexor carpi ulnaris and medial to the ulnar artery, before giving off the dorsal branch of the ulnar 6 cm proximal to the wrist. (The dorsal branch becomes rapidly subcutaneous and supplies the skin on the ulnar posterior aspect of the hand.)
Passes into the hand superficial to the flexor retinaculum but deep to the palmar carpal ligament (which runs between the pisiform bone and the

hook of hamate, both of which are easily palpable). This space is known as Guyon's canal.
Continues as digital nerves along the digits.

- Motor – in the wrist, flexor carpi ulnaris and half of flexor digitorum profundus. In the hand, all of the intrinsic muscles not served by the median nerve (all intrinsic muscles except 'LOAF').
- Sensory – medial 1.5 digits (digital nerves) and medial hand anteriorly and posteriorly.

vi) Medial cutaneous nerve of the arm; medial cutaneous nerve of the forearm

Both arise from the medial cord of the brachial plexus (Figure 4.6) to supply the skin of the arm and forearm respectively.

vii) Intercostobrachial nerve

This is a cutaneous branch of the second intercostal nerve (from T2) and supplies the skin of the medial arm.

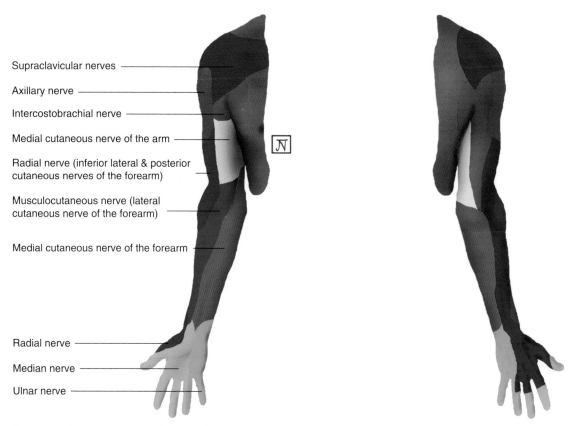

Supraclavicular nerves

Axillary nerve

Intercostobrachial nerve

Medial cutaneous nerve of the arm

Radial nerve (inferior lateral & posterior cutaneous nerves of the forearm)

Musculocutaneous nerve (lateral cutaneous nerve of the forearm)

Medial cutaneous nerve of the forearm

Radial nerve

Median nerve

Ulnar nerve

Figure 6.5 Cutaneous nerves of the upper limb. The supraclavicular nerves originate in the superficial cervical plexus and the intercostobrachial nerves originate from the second intercostal nerve (T2); all other nerves originate from the brachial plexus.

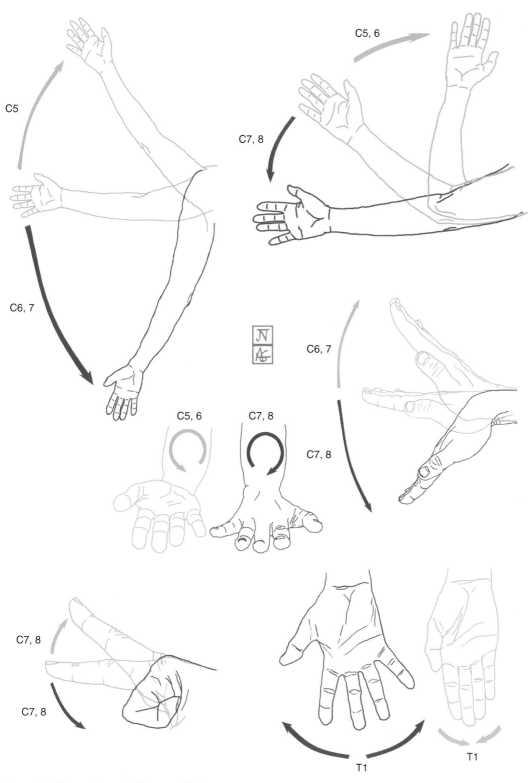

C5

C6, 7

C5, 6

C7, 8

C7, 8

C5, 6

C7, 8

C6, 7

C7, 8

C7, 8

C7, 8

T1

T1

Figure 6.6 The myotomes of the upper limb.

Selective nerve blocks at the elbow

Introduction

Effective block requires anaesthetising the radial, median and ulnar nerves as well as the lateral cutaneous nerve of the forearm (continuation of the musculocutaneous nerve) and the medial cutaneous nerve of the forearm (a direct branch of the medial cord of the brachial plexus). Refer to Figures 6.7 and 6.8.

Indications

Hand surgery where forearm motor block is required but proximal brachial plexus block is not desired.

Specific contraindications

- Care should be taken if the elbow block is used to supplement an incomplete proximal brachial plexus block, as the patient's perception of neurological injury may be attenuated.[1]

Technique

Landmark/nerve stimulator

Radial nerve
The arm is supinated and abducted. 3–5 ml of local anaesthetic is injected 1.5 cm lateral to the biceps tendon at a depth of 2–4 cm. Stimulation produces thumb extension.

Median nerve
3–5 ml of local anaesthetic is injected just medial to the brachial artery at the level of the epicondyles. Stimulation produces flexion of the first three digits.

Ulnar nerve
The forearm is flexed at the elbow and 2–3 ml of local anaesthetic is injected 2 cm proximal to the medial epicondyle at 45° to the skin in a proximal direction. Stimulation produces thumb adduction and flexion of the fifth digit. The volume is limited to avoid excessive pressure within the tight fascial space of the ulnar groove and consequent compromise of neural blood flow.

Lateral cutaneous nerve of the forearm (LCNF)
The LCNF is blocked by injecting 5 ml of local anaesthetic deep to the lateral margin of the biceps tendon at the level of the epicondyles, and a further 5 ml subcutaneously laterally along the elbow crease. The LCNF is sensory only.

Medial cutaneous nerve of forearm (MCNF)
The MCNF is blocked by injecting 5 ml subcutaneously in a half-ring block above the medial epicondyle. The MCNF is sensory only.

Ultrasound-assisted block

Radial, median and ulnar nerves
The radial, median and ulnar nerves are blocked as described in the ultrasound section of the wrist block (see box on *Wrist block*).

Lateral cutaneous nerve of the forearm (LCNF)
The musculocutaneous nerve can be scanned with a high-frequency probe from the axilla (see box on *Axillary brachial plexus block*) down the arm, where it passes under the biceps muscle before passing laterally to become the LCNF, medial and adjacent to the cephalic vein.
Light pressure with the probe enables the cephalic vein to be seen. A 50 mm needle is introduced in-plane and 2 ml of local anaesthetic is injected medial to the cephalic vein.

Medial cutaneous nerve of forearm (MCNF)
The arm is abducted to 90° and externally rotated. A high-frequency probe is placed on the medial side of the elbow and the basilic vein is identified with the two branches (anterior and posterior) of the MCNF around it. The basilic vein is traced proximally until the nerve branches unite, at which point 2 ml of local anaesthetic is injected in-plane with a 50 mm needle.[2]

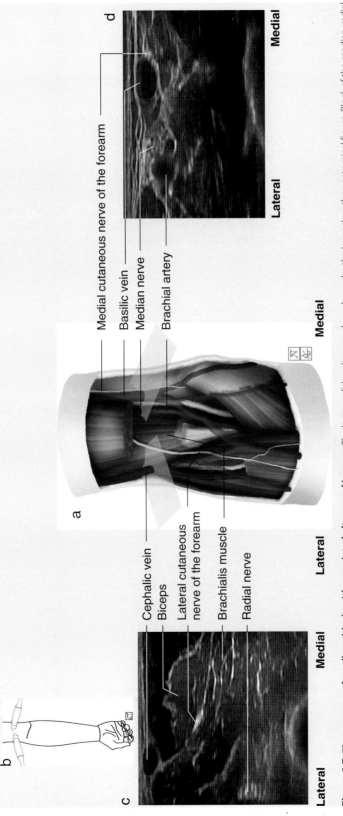

Medial cutaneous nerve of the forearm
Basilic vein
Median nerve
Brachial artery

Cephalic vein
Biceps
Lateral cutaneous nerve of the forearm
Brachialis muscle
Radial nerve

Medial

Lateral

Medial

Lateral

Lateral

Medial

Figure 6.7 The anatomy of an elbow block, with associated ultrasound images. The beams of the ultrasound are shown by the 'panes' on the anatomical figure. Block of the median, radial and ulnar nerves is described in the *Wrist block* box. A segment of biceps brachii has been removed to reveal the musculocutaneous nerve, and likewise a segment of brachioradialis to reveal the superficial branch of the radial nerve. Note that the lateral cutaneous nerve of the forearm is the continuation of the musculocutaneous nerve; the medial cutaneous nerve of the forearm arises directly from the brachial plexus.

Complications

• Residual paraesthesia due to inadvertent intraneural injection.

References

1. Neal J, Hebl J, Gerancher J, *et al*. Brachial plexus anesthesia: essentials of our current understanding. *Reg Anesth Pain Med* 2002; **27**: 402–28.
2. Thallaj A. Ultrasound guidance of uncommon nerve blocks. *Saudi J Anesth* 2011; **5**: 392–4.

Wrist block

Introduction

Wrist block produces anaesthesia of the hand. It may be used for awake surgery or in combination with general anaesthesia. It may also be used in combination with short-acting local anaesthetic agents in a brachial plexus block, providing prolonged pain relief once the more proximal block has worn off.

Indications

Most commonly used for carpal tunnel, hand and finger surgery.

Specific contraindications

• Local infection.
• Prolonged surgery – patients can usually tolerate an arm tourniquet for about 20 minutes or a wrist tourniquet for about 120 minutes.

There is a commonly held belief that adrenaline causes irreversible vasospasm and digital infarction, based on case reports from the 1950s using procaine or cocaine.[1] However, a multicentre prospective study in 2005 suggested that use of lidocaine and low-dose adrenaline (1:100,000) in well-vascularised hands and fingers has a low risk of digital ischaemia and infarction.[2]

Technique

Landmark/nerve stimulator

Ulnar nerve

The wrist is placed in a supine position with slight dorsiflexion and the arm abducted. A 25G needle is inserted from the medial side of the wrist under the tendon of flexor carpi ulnaris, just anterior to the ulnar styloid process. 3–5 ml of local anaesthetic is injected just inferior and 5–10 mm lateral to the tendon. If blood is aspirated then the needle is redirected more superficially and medially.

An alternative approach is to block the nerve in Guyon's canal (see text). The pisiform and the hook of hamate are palpated on the medial side of the wrist and the needle is inserted 1 cm proximally, at 45° to the skin and directed towards the canal.

Ulnar nerve stimulation produces paraesthesia in the fifth digit.

Dorsal branch of the ulnar nerve

The dorsal branch (which supplies sensation to the ulnar aspect of the back of the hand – see text) arises from the medial aspect of the ulnar nerve and becomes subcutaneous 5 cm proximal to the pisiform.[3] The forearm is pronated, the needle is inserted 4 cm proximal to the ulnar styloid, and 3–5 ml of local anaesthetic is injected as a subcutaneous wheal around the medial aspect of the ulnar.

Median nerve

The wrist is placed in a supine position. The needle is inserted at 45° to the skin aiming towards the hand, at a point 2.5 cm proximal to the wrist crease, between the tendons of palmaris longus and flexor carpi radialis. It is advanced a few millimetres deep to the plane of the tendons, and 3–5 ml of local anaesthetic is injected. 2 ml of local anaesthetic is then injected in a 30° subcutaneous fan, ensuring capture of the palmar cutaneous branch – see text.
Median nerve stimulation produces paraesthesia in the first three digits.

Radial nerve

The radial nerve divides into multiple small cutaneous branches about 5 cm proximal to the radial styloid. The needle is inserted 3 cm proximal to the radial styloid and a field block of 3–5 ml of local anaesthetic is injected subcutaneously around the radius, towards the midpoint of the dorsum of the wrist.

Posterior interosseous nerve

The posterior interosseous nerve is a continuation of the deep branch of the radial nerve. The forearm is pronated and the soft spot palpated in the midline 2 cm proximal to the distal radioulnar joint. The needle is inserted perpendicularly to the skin until resistance is felt as the interosseous membrane is reached, then withdrawn slightly, and 2 ml of local anaesthetic is injected.

Anterior interosseous nerve

The anterior interosseous nerve is a branch of the median nerve. From the position above, the needle is advanced further through the interosseous membrane and a 'click' may be felt. 2 ml of local anaesthetic is injected.

Ultrasound-assisted block[4]

Ulnar nerve

The block is performed 2 cm proximal to the point where the ulnar nerve and ulnar artery converge (Figure 6.8e).[5] Block at this level should also provide anaesthesia of the dorsal branch, which usually arises about 6 cm proximal to the wrist.
Distally the ulnar nerve lies superficial and medial to the ulnar artery all the way to the wrist and appears hypoechoic with a fine fascicular honeycomb pattern.

Radial nerve

For surgery involving the wrist, the radial nerve is blocked about 5 cm proximal to the lateral epicondyle (Figure 6.8b), proximal to the emergence of the posterior interosseous nerve (deep branch of the radial nerve). At this level the radial nerve is easily identified,[6] as the surrounding muscles form a clear acoustic window,[7] and both branches are blocked with a single injection. The radial nerve may also be easily identified lateral to the radial artery and blocked more distally if wrist analgesia is not required (Figure 6.8a).

Median nerve

For surgery involving the wrist the median nerve is blocked in the antecubital fossa, proximal to the emergence of the anterior interosseous nerve (Figure 6.8d),[8] where it appears as a hyperechoic structure medial to the brachial artery. The median nerve may also be easily identified in the midline of the forearm, avoiding the brachial artery, if wrist analgesia is not required.

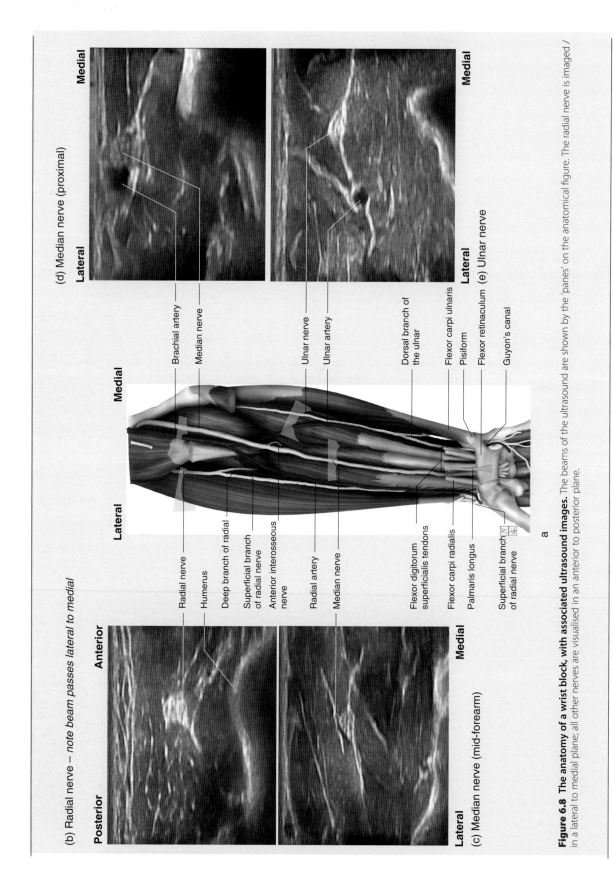

(b) Radial nerve – *note beam passes lateral to medial*

Posterior

Anterior

(c) Median nerve (mid-forearm)

(d) Median nerve (proximal)

Lateral

Medial

(e) Ulnar nerve

Lateral

Medial

Lateral

Medial

Radial nerve

Humerus

Deep branch of radial

Superficial branch of radial nerve

Anterior interosseous nerve

Radial artery

Median nerve

Flexor digitorum superficialis tendons

Flexor carpi radialis

Palmaris longus

Superficial branch of radial nerve

Brachial artery

Median nerve

Ulnar nerve

Ulnar artery

Dorsal branch of the ulnar

Flexor carpi ulnaris

Pisiform

Flexor retinaculum

Guyon's canal

Lateral

Medial

a

Figure 6.8 The anatomy of a wrist block, with associated ultrasound images. The beams of the ultrasound are shown by the 'panes' on the anatomical figure. The radial nerve is imaged / in a lateral to medial plane; all other nerves are visualised in an anterior to posterior plane.

Complications

- Residual paraesthesia due to inadvertent intraneural injection.

References

1. Denkler K. A comprehensive review of epinephrine in the finger: to do or not do. *Plast Reconstr Surg* 2001; **108**: 114–24.
2. Lalonde D, Bell M, Sparkes G, *et al*. A multicenter prospective study of 3,110 consecutive cases of elective epinephrine use in the fingers and hand: the Dalhousie Project clinical phase. *J Hand Surg* 2005; **30**: 1061–7.
3. Botte M, Cohen M, Lavernia C, *et al*. The dorsal branch of the ulnar nerve: an anatomic study. *J Hand Surg* 1990; **15**: 603–7.
4. Colin J, McCartney M, Daquan X, *et al*. Ultrasound examination of peripheral nerves in the forearm. *Reg Anesth Pain Med* 2007; **32**: 434–9.
5. Kathirgamanathan A, French J, Foxall G, *et al*. Ultrasound anatomy of the ulnar nerves in the upper and forearm: 171 *Reg Anesth Pain Med* 2007; **32**: 3 (Free papers).
6. Retzl G, Kapral S, Greher M, *et al*. Ultrasonographic findings of the axillary part of the brachial plexus. *Anesth Analg* 2001; **92**: 1271–5.
7. Foxall G, Skinner D, Hardman J, *et al*. Ultrasound anatomy of the radial nerve in the distal upper arm. *Reg Anesth Pain Med* 2007; **32**: 217–20.
8. Frazao R, Alves N, Cricenti S. The origin and point of penetration of the nerve branches supplying the flexor digitorum profundus. *Braz J Morphol Sci* 2000; **17**: 113–16.

Digital block

Introduction

Digital block provides anaesthesia of the fingers or toes. It is simple to perform and has few systemic complications.

Indications

Used for a range of minor surgical procedures on the digits such as lacerations of the finger or toe, nail-bed repair, removal of foreign bodies and paronychia drainage.

Specific contraindications

- Infection in the digit.
- Compromised circulation in the digit.
- Compromised nerve function.

There is a commonly held belief that adrenaline causes irreversible vasospasm and digital infarction, based on case reports from the 1950s using procaine or cocaine.[1] However, a multicentre prospective study in 2005 suggested that use of lidocaine and low-dose adrenaline (1:100,000) in well-vascularised hands and fingers has a low risk of digital ischaemia and infarction.[2]

Technique

The injection volume should be limited to 3 ml on each side, to reduce the mechanical pressure effect of the local anaesthetic in the confined space at the base of the digit.

It may be advantageous to combine the web space and flexor sheath approaches described below.

Landmark

Web space block

The patient's hand is placed palm down on the sterile field. A 25G needle is inserted perpendicularly just distal to the metacarpal–phalangeal joint and directed towards the base of the phalanx. The needle should not penetrate the palmar dermis opposite the needle path. 2–3 ml of local anaesthetic is injected in the anterior aspect of the web space and an additional 1 ml is injected continuously as the needle is withdrawn. The procedure is repeated on each side of the finger being anaesthetised.

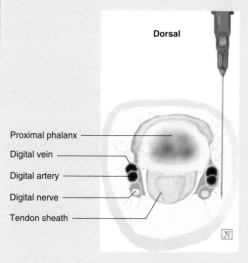

Dorsal

Proximal phalanx

Digital vein

Digital artery

Digital nerve

Tendon sheath

Palmar

Figure 6.9 Cross-section of the proximal phalanx illustrating a web space (digital nerve) block.

Flexor sheath block[3]

This technique produces anaesthesia by a single injection into the flexor tendon sheath at the level of the distal palmar crease. The local anaesthetic spreads within the sheath and diffuses circumferentially around the proximal phalanx.

The patient's hand is placed supine on the sterile field and the flexor tendon sheath is palpated at the distal palmar crease. A 25G needle is inserted at 45° to the palmar skin, pointing towards the fingers, and advanced into the sheath. 2–3 ml of local anaesthetic is injected. If resistance is felt then the needle is against the flexor tendon and should be withdrawn. Proximal pressure may promote distal diffusion of the local anaesthetic. The onset time of the block is about 5 minutes.

An alternative approach is to insert the needle at 90° at the metacarpal crease until bone is contacted and then withdraw slightly before injection.

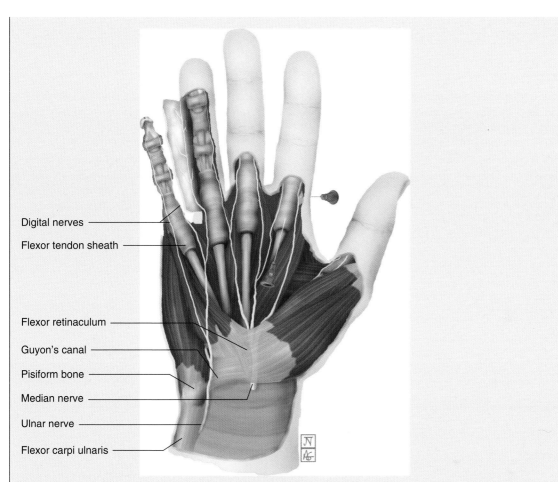

Digital nerves

Flexor tendon sheath

Flexor retinaculum

Guyon's canal

Pisiform bone

Median nerve

Ulnar nerve

Flexor carpi ulnaris

Figure 6.10 The anatomy of a web space block and a flexor sheath block. Note how the ulnar nerve supplies the medial 1.5 digits; the median nerve supplies the remainder of the cutaneous sensation to the hand anteriorly. The upper needle depicts a web space block; the lower needle depicts a flexor sheath block.

Complications

- Gangrene of the digit – see contraindications (injection volume, adrenaline and small-vessel disease).
- Nerve injury (rare).

References

1. Denkler K. A comprehensive review of epinephrine in the finger: to do or not do. *Plast Reconstr Surg* 2001; **108**: 114–24.
2. Lalonde D, Bell M, Sparkes G, *et al*. A multicenter prospective study of 3,110 consecutive cases of elective epinephrine use in the fingers and hand: the Dalhousie Project clinical phase. *J Hand Surg* 2005; **30**: 1061–7.
3. Chiu D Transthecal digital block: flexor tendon sheath used for anaesthetic infusion. *J Hand Surg* 1990; **15**: 471–3.

4. The cubital fossa

A triangular area on the anterior aspect of the elbow.

i) Boundaries

- Superior – an imaginary line connecting the medial and lateral epicondyles.
- Medial – lateral border of pronator teres.
- Lateral – medial border of brachioradialis.
- Floor – brachialis and supinator muscles.
- Roof – bicipital aponeurosis and deep fascia.

ii) Contents

- Arteries – brachial and its terminal branches, the radial and ulnar arteries.
- Vein – brachial vein.
- Nerves – median, radial (deep and superficial branches).
- Tendon of biceps.

Overlying the fossa are the median cubital vein and the medial and lateral cutaneous nerves of the forearm. Mnemonic (from medial to lateral) – Mary's Big Toe Protrudes Rather (Median nerve, Brachial artery, Tendon of biceps, Perforating vein, Radial nerve).

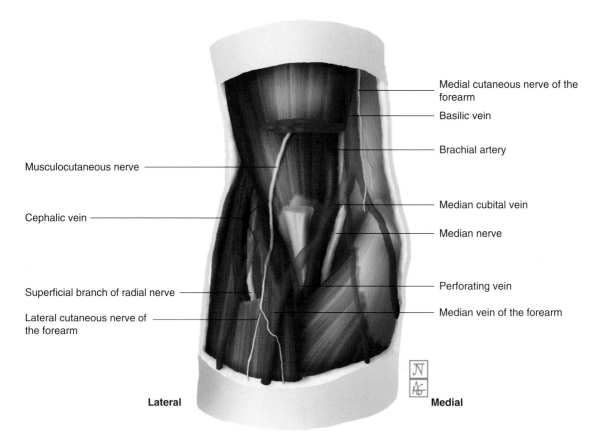

Medial cutaneous nerve of the forearm

Basilic vein

Brachial artery

Median cubital vein

Median nerve

Perforating vein

Median vein of the forearm

Musculocutaneous nerve

Cephalic vein

Superficial branch of radial nerve

Lateral cutaneous nerve of the forearm

Lateral

Medial

Figure 6.11 The cubital fossa of the right arm. A segment of biceps brachii has been removed to reveal the musculocutaneous nerve, and likewise a segment of brachioradialis to reveal the superficial branch of the radial nerve. Note that the lateral cutaneous nerve of the forearm is the continuation of the musculocutaneous nerve; the medial cutaneous nerve of the forearm arises directly from the brachial plexus.

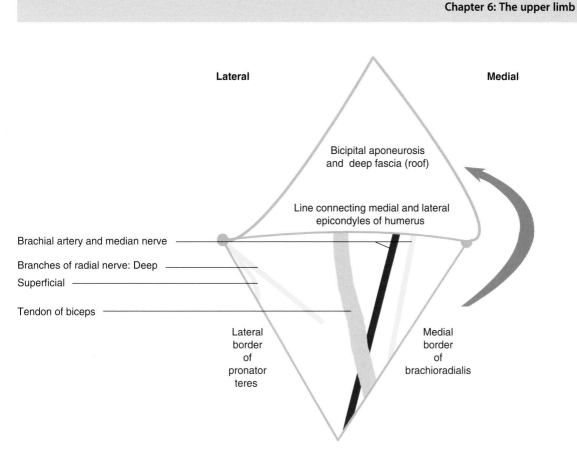

Figure 6.12 Schematic illustration of the right cubital fossa. The floor of the fossa is formed from brachialis and supinator muscles. Mnemonic (from medial to lateral) – 'Mary's Big Toe Protrudes Rather' Median nerve, Brachial artery, Tendon of biceps, Perforating vein (not shown), Radial nerve.

Biers block

Introduction

Intravenous regional anaesthesia (IVRA) is performed by administering intravenous local anaesthetic into a limb that has been isolated from the circulation by a cuff. It is a simple, effective technique which is safe and effective when performed properly.

Indications

Operations distal to the elbow or knee that take under 40 minutes, e.g. fracture reduction. After 40–60 minutes tourniquet pain develops, which limits its use. Use of a double-cuffed tourniquet may reduce this discomfort (the upper cuff is inflated first, and then it is switched to the lower cuff). The use of intravenous adjunctive analgesia such as fentanyl may also be of value.

Specific contraindications

- Raynaud's disease.
- Sickle cell disease.
- Crush injury.
- Young children (poorly tolerated unless sedated).
- Unstarved patient – as sedation or GA may be required.

- Bupivacaine is unsuitable because of its cardiotoxic profile.
- Adrenaline-containing local anaesthetic solutions should be avoided.

Pre-procedure checks

See *Universal procedural advice* – in particular, full resuscitation equipment, ECG monitoring and separate IV access to the circulation for use in the event of complications.
The single or double tourniquet must be checked to ensure it does not leak.

Technique

Prilocaine is the preferred local anaesthetic agent (maximum dose 6 mg/kg), using 40 ml 0.5% prilocaine for the upper limb and 60 ml for the lower limb. Lidocaine is a useful alternative (maximum dose 3 mg/kg).

Landmark

A cannula is inserted in the non-operative limb. A second cannula is inserted as distal as possible in the operative limb. A baseline blood pressure is measured and the occlusive tourniquet is placed on the upper part of the operative arm with wadding. The operative limb is exsanguinated by elevation or with an Esmarch bandage and the tourniquet is inflated to 50–100 mmHg above the systolic blood pressure. It is important to check that pulses are absent in the limb and that the cuff maintains a continuous pressure.
Local anaesthetic is injected slowly into the operative limb, and the onset of anaesthesia begins after about 5 minutes. The tourniquet must not be deflated until at least 20 minutes has elapsed after injection.
After the procedure the IVRA cannula and tourniquet are removed.

Complications

- Accidental leak or deflation of tourniquet – dizziness, nausea, vomiting, tinnitus, perioral tingling, muscle twitching, loss of consciousness, convulsions and death.

Post-procedure checks

- The patient must be observed with BP and ECG monitoring for a minimum of 10 minutes after cuff deflation.

The abdomen

1. Landmarks of the abdomen

The surface landmarks of the abdomen are:
- Superiorly – the xiphoid process (level T9) and the costal margin.
- Inferiorly – the inguinal ligaments and pubic symphysis.

The umbilicus is found at level L3/4.
A line joining the iliac crests passes through the L4–5 interspace (Tuffier's line).

2. Layers

From superficial to deep:
- Skin.
- Camper's fascia.
- Scarpa's fascia.
- External oblique muscle.
- Internal oblique muscle.
- Transverse abdominal muscle.
- Transversalis fascia.
- Extraperitoneal fat.
- Parietal peritoneum.

3. Muscles

i) Rectus abdominis

Two paired muscles.
Origin: pubic symphysis and crest. Insertion: xiphoid process and fifth–seventh costal cartilages.
Surrounded by the rectus sheath:
- The sheath is formed by the aponeurosis of internal oblique, which splits to join the aponeurosis of external oblique anteriorly and the aponeurosis of transverse abdominal posteriorly.
- The sheath is deficient superior to the costal margin and inferior to the arcuate line (one-third of the distance from the umbilicus to the pubic crest). Above the costal margin, rectus abdominis lies directly on the thoracic wall. Below the arcuate line, the aponeuroses of all three of the flat abdominal muscles pass anterior to rectus abdominis. A rectus sheath block must therefore be performed between the costal margin and the arcuate line (see box on *Rectus sheath block*, below).
- The sheath contains rectus abdominis, the superior and inferior epigastric vessels, the subcostal vessels and the terminal branches of the ventral rami of T7–12.

Three transverse tendinous intersections (at the level of the xiphoid, the umbilicus and halfway between the two) anchor rectus to the anterior rectus sheath, precluding effective spread of local anaesthetic within the anterior rectus sheath. For this reason, a rectus sheath block is performed between rectus and the posterior rectus sheath, where no such obstructions exist.
Between the rectus muscles is the linea alba, the site of fusion of the two oblique and the transverse abdominal muscles. It is a useful reference point when performing ultrasound-guided blocks of the abdominal wall.

ii) External oblique

Most superficial of the three flat abdominal muscles, and continuous with the external intercostal muscles.
Fibres run inferomedially ('hands in pockets'), becoming aponeurotic in the mid-clavicular line.

iii) Internal oblique

Deep to external oblique and continuous with the internal intercostal muscles.
Fibres run superomedially and become aponeurotic in the mid-clavicular line.

iv) Transverse abdominal

Innermost muscle, continuous with transversus thoracis muscles.
Fibres run horizontally and again become aponeurotic medially.

Rectus sheath block

Introduction

Injection of local anaesthetic bilaterally between the rectus muscle and the posterior rectus sheath provides dense analgesia over the middle anterior abdominal wall for structures superficial to the peritoneum. In up to 30% of the population the anterior cutaneous branches are formed before the rectus sheath, leading to block failure.[1]

The block is often performed after induction of anaesthesia and before surgical incision, but it may also be performed as a rescue block in an awake or sedated patient.

Indications

Midline laparotomy,[2] reducing the opioid requirements for somatic pain. It is particularly useful for surgery around the umbilical area.

Specific contraindications

- The block cannot be performed inferior to the arcuate line, because of deficiency of the posterior rectus sheath.

Technique

Landmark

The patient is placed in a supine position and a short-bevelled 5 cm needle is introduced at right angles to the skin 2–3 cm from the midline, slightly cephalad to the umbilicus. A 'pop' is felt as the needle passes through the anterior rectus sheath, and the needle is advanced through the rectus abdominis muscle until the firm resistance of the posterior wall is felt. 15–20 ml of local anaesthetic is injected in 5 ml aliquots and then the procedure is repeated on the contralateral side. The landmark approach is a blind technique using tactile stimuli, and it should be noted that there is poor correlation between depth and body habitus.

Ultrasound-assisted block

A high-frequency probe is placed in a transverse plane above or below the umbilicus and the linea alba is identified. The probe is moved laterally until the linea semilunaris is identified at the lateral edge of the rectus abdominis muscle. The block needle is inserted medial to the linea semilunaris either in-plane or out-of-plane. A perpendicular out-of-plane insertion may improve appreciation of the needle passage through the various layers.

The anterior rectus sheath is identified on ultrasound, and movement of the needle back and forth along the sheath may elicit a scratching sensation. The needle is advanced further through the sheath and into the muscle belly. The posterior aspect of the rectus sheath appears as a hyperechoic twin line and may again be appreciated by a scratching sensation. Hydrolocation (visualising the needle tip by identifying an area of expanding hypoechogenicity following a small injection of local anaesthetic) is useful to confirm correct positioning. 20 ml of 0.25% bupivacaine hydrodissects the rectus muscle away from the posterior rectus sheath and produces about 6 hours of analgesia. Resistance to injection may be due to injection into the body of the muscle and requires repositioning.

A rectus sheath catheter may be placed using a similar technique, but the probe is placed longitudinally and an 18G Tuohy needle is introduced in-plane at 45° to the skin. Following an initial bolus (as above) to open the space, the catheter is fed about 8 cm into the space and a dressing is applied away from the surgical field. The technique is repeated on the opposite side. Placement via the anterior abdominal wall using ultrasound has potential advantages over surgical placement in that it enables intraoperative analgesia, prevents loss of local anaesthetic into the abdominal cavity, and maintains isolation of the sterile rectus sheath compartment.

Complications

- Injection into peritoneal cavity leading to block failure and puncture of bowel.
- Puncture of the vessels within the sheath (see text) leading to intravascular injection or haematoma.
- Injection directly into the rectus muscle may increase the risk of systemic absorption.

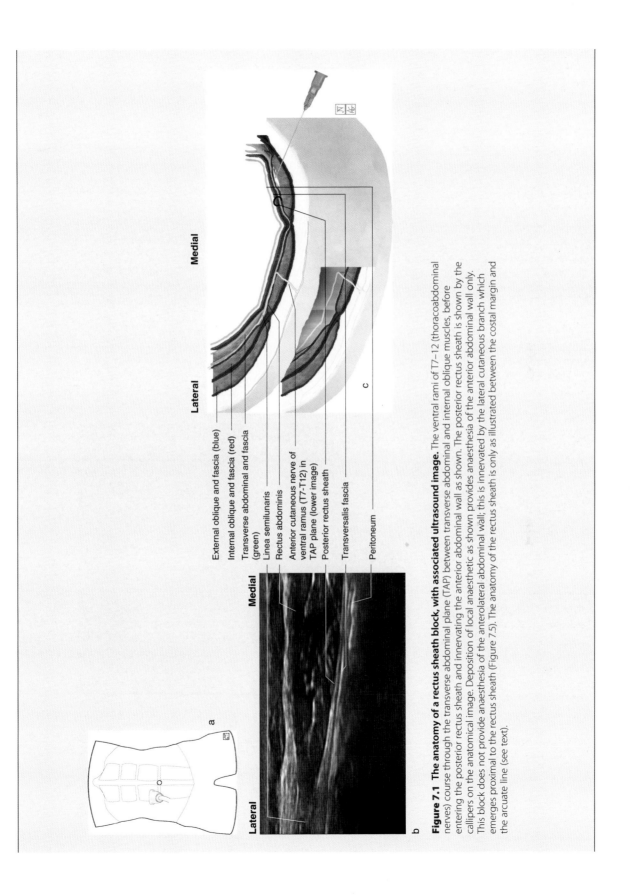

Figure 7.1 The anatomy of a rectus sheath block, with associated ultrasound image. The ventral rami of T7–12 (thoracoabdominal nerves) course through the transverse abdominal plane (TAP) between transverse abdominal and internal oblique muscles, before entering the posterior rectus sheath and innervating the anterior abdominal wall as shown. The posterior rectus sheath is shown by the callipers on the anatomical image. Deposition of local anaesthetic as shown provides anaesthesia of the anterior abdominal wall only. This block does not provide anaesthesia of the anterolateral abdominal wall; this is innervated by the lateral cutaneous branch which emerges proximal to the rectus sheath (Figure 7.5). The anatomy of the rectus sheath is only as illustrated between the costal margin and the arcuate line (see text).

Lateral

Medial

a

b

Lateral

Medial

External oblique and fascia (blue)
Internal oblique and fascia (red)
Transverse abdominal and fascia (green)
Linea semilunaris
Rectus abdominis
Anterior cutaneous nerve of ventral ramus (T7–T12) in TAP plane (lower image)
Posterior rectus sheath
Transversalis fascia
Peritoneum

c

References

1. Skinner A, Lauder G. Rectus sheath block: successful use in the chronic pain management of pediatric abdominal wall pain. *Paediatr Anaesth* 2007; **17**: 1203–11.
2. Webster K, Hubble S. Rectus sheath analgesia in intensive care patients: technique description and case series. *Anaesth Intens Care* 2009; **37**: 855.

4. The lumbar plexus

Formed from the ventral rami of L1–4 (in 50% may be T12–L5).
Supplies:

- Motor and sensory innervation to the lower abdomen, scrotum or labia majora, and lower limb.

The plexus is formed in front of the transverse processes of L1–4 within the substance of psoas major muscle.
Terminal nerves:

- Ilioinguinal and iliohypogastric nerves (L1). See section 6, below.
- Genitofemoral nerve (L1, 2). Divides into genital and femoral branches:

The femoral branch supplies the skin below the inguinal ligament.
The genital branch in the male passes through the inguinal canal in the spermatic cord to supply cremaster muscle and the skin of the scrotum; in the female it accompanies the round ligament of the uterus to supply the labium majus and the mons pubis.

- Lateral cutaneous nerve of the thigh (L2, 3).
- Femoral nerve (L2, 3, 4).
- Obturator nerve (L2, 3, 4).

These last three nerves and the blocks relating to them are described in more detail in Chapter 8.

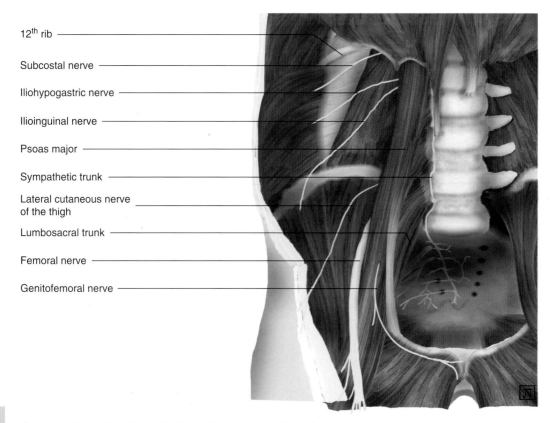

12th rib
Subcostal nerve
Iliohypogastric nerve
Ilioinguinal nerve
Psoas major
Sympathetic trunk
Lateral cutaneous nerve of the thigh
Lumbosacral trunk
Femoral nerve
Genitofemoral nerve

Figure 7.2 The lumbar plexus. The plexus is formed within the body of psoas muscle. The obturator nerve lies posterior to psoas major muscle at the pelvic brim so is not seen on this image.

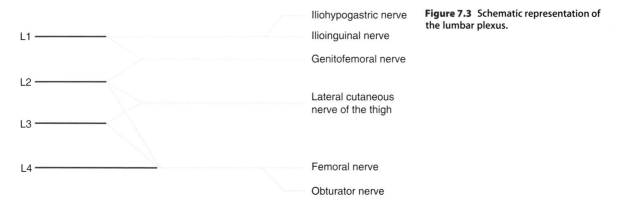

Figure 7.3 Schematic representation of the lumbar plexus.

Lumbar plexus block

Introduction

Lumbar plexus block provides anaesthesia for lower limb surgery by blocking the lateral cutaneous nerve of the thigh, femoral, obturator and genitofemoral nerves. A successful block requires spread of a large volume of local anaesthetic within the psoas muscle in the fascial plane where the roots of the plexus are situated.
Lumbar plexus block has the potential for serious complications, such as systemic absorption and epidural or spinal spread.[1]

Indications

Fractures of the femoral neck or shaft, knee arthroscopy and anterior thigh procedures.

Specific contraindications

- Clopidogrel – death from a massive retroperitoneal haematoma following lumbar plexus block has been described.[2]

Pre-procedure checks

Sedation may be indicated.

Technique

The local anaesthetic commonly spreads cranially around the femoral nerve distribution (L2–4). In over 50% of patients the obturator nerve is separated from the femoral nerve fibres by a muscle fold,[3] so a relatively large volume of local anaesthetic is required to spread to the entire plexus. There is significant potential for inadvertent systemic absorption and epidural spread so a slow, gentle injection and use of adrenaline as a vascular marker is suggested.

Landmark/nerve stimulator

The patient is in the lateral position with the operative leg uppermost, and hips and knees flexed to 90°. A line is drawn laterally from the centre of the L4 spinous process, terminating at a vertical parasagittal line passing through the posterior superior iliac spine (PSIS). The needle insertion point is at the junction of the lateral third and medial two-thirds of this line (about 3–4 cm lateral to the L4 spinous process).[4] The needle is advanced at right angles to the skin until the transverse process of L4 is encountered. It is then redirected caudally until quadriceps twitches are obtained at approximately 6–8 cm depth, following which 25–35 ml of local anaesthetic is injected slowly with frequent aspiration.
Twitches in the hamstrings suggest that the needle insertion point is too caudal, producing stimulation of the sciatic plexus. Needle insertion to 10 cm with no bony contact or twitches suggests placement is too lateral and has missed the nerve roots.

Ultrasound-assisted block

Ultrasound imaging of the lumbar plexus is technically difficult because of its depth, and the optimal approach has not yet been established. For this reason the authors recommend simultaneous use of a nerve stimulator.

Paramedian transverse scan[5]

The patient is placed in the lateral position with the hip and knees slightly flexed. The vertebral level is identified (see box on *Epidural anaesthesia*) and a transverse image of the L4 transverse process is obtained. The probe is slid inferiorly until the acoustic shadow of the transverse process disappears. The beam is directed medially to identify the psoas muscle deep to the erector spinae and adjacent to the vertebral bodies. The nerve roots commonly lie between the anterior two-thirds and posterior third of the psoas muscle at an average depth of 5–6 cm.[6] Colour Doppler is useful to identify the adjacent vascular structures. The needle is inserted in-plane but may be poorly visualised because of the steep angle of insertion. Psoas twitches may be seen as the needle enters the muscle, but injection is not performed until quadriceps contraction is observed.

Parasagittal scan

The patient is positioned as above and the low-frequency probe is placed in a parasagittal plane about 3 cm lateral to the midline. The probe is positioned over the transverse processes of L2, L3 and L4 by counting up from the lumbosacral junction. The psoas muscle may then be seen in the acoustic window between the transverse processes with the hyperechoic nerve roots within it.[7] An in-plane technique with nerve stimulation is used, as described above.

Complications

The proximity of the nerve root dural cuffs risks epidural or spinal anaesthesia (the former in 15% of patients) leading to hypotension.[8]
- Local anaesthetic toxicity – higher risk than most other nerve blocks due to high-volume injection into deep well-vascularised muscle.
- Nerve injury – uncommon.
- Vascular puncture – uncommon, but deep needle insertion may puncture the vena cava or aorta.
- Iliopsoas or renal haematoma – particularly if multiple needle passes.

Post-procedure checks

- Postoperative observations as for epidural anaesthesia, to allow early detection of inadvertent high neuraxial block.

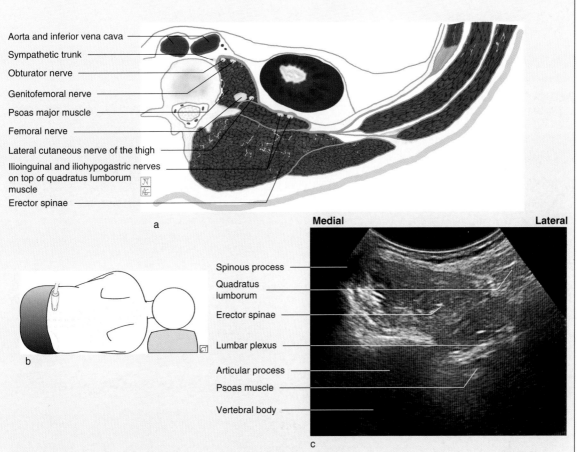

Aorta and inferior vena cava
Sympathetic trunk
Obturator nerve
Genitofemoral nerve
Psoas major muscle
Femoral nerve
Lateral cutaneous nerve of the thigh
Ilioinguinal and iliohypogastric nerves
on top of quadratus lumborum
muscle
Erector spinae

a

Medial   Lateral

Spinous process
Quadratus
lumborum
Erector spinae

Lumbar plexus

Articular process
Psoas muscle

Vertebral body

b

c

Figure 7.4 The lumbar plexus with associated ultrasound image. (a) Transverse section through L4 (Tuffier's line), viewed from above, showing the nerves of the lumbar plexus. The nerves form within the body of psoas major muscle; as the ilioinguinal and iliohypogastric nerves are formed high within the muscle, they can be seen to have exited psoas and be travelling over quadratus lumborum muscle at this level. (b, c) Ultrasound image of the lumbar plexus.

References

1. Auroy Y, Benhamou D, Bargues L, et al. Major complications of regional anesthesia in France. *Anesthesiology* 2002; **97**: 1274–80.
2. Maier C, Gleim M, Weiss T, et al. Severe bleeding following lumbar sympathetic blockade in two patients under medication with irreversible platelet aggregation inhibitors. *Anesthesiology* 2002; **97**: 740–3.
3. Sim I, Webb T. Anatomy and anaesthesia of the lumbar somatic plexus. *Anaesth Intensive Care* 2004; **32**: 178–87.
4. Capdevila X, Macaire P, Dadure C, et al. Continuous psoas compartment block for postoperative analgesia after total hip arthroplasty: new landmarks, technical guidelines and clinical evaluation. *Anesth Analg* 2002; **94**: 1606–13.
5. Karmakar M, Li J, Kwok W, et al. Sonoanatomy relevant for lumbar plexus block in volunteers correlated with cross-sectional anatomic and magnetic resonance images. *Reg Anesth Pain Med* 2013; **38**: 391–7.
6. Farny J, Drolet P, Girard M. Anatomy of the posterior approach to the lumbar plexus block. *Can J Anesth* 1994; **41**: 480–5.
7. Karmakar M, Ho A, Li X, et al. Ultrasound-guided lumbar plexus block through the acoustic window of the lumbar ultrasound trident. *Br J Anaesth* 2008; **100**: 533–7.
8. Molina M, Asensio R, Barrio M, et al. Epidural anaesthesia after posterior lumbar plexus block. *Rev Esp Anestesiol Reanim* 2005; **52**: 55–6

5. Neurovascular supply

i) Arterial

Predominantly derived from the superior and inferior epigastric arteries (derived from the internal thoracic and external iliac arteries respectively). These vessels run within the rectus sheath.

ii) Venous

Corresponding veins.

iii) Nervous

A neurovascular plane exists between the muscles of internal oblique and transverse abdominal, known as the transverse abdominal plane or TAP. It corresponds to a similar plane in the thorax (Figure 5.8). The nerves within this plane supply the muscles of the anterior and lateral abdominal wall and are:

- Terminal branches of ventral rami of nerves T7–12 (thoracoabdominal nerves).
- Iliohypogastric and ilioinguinal nerves (derived from L1).

The nerves course through the plane in an anteroinferior manner, prior to entering the posterior rectus sheath and emerging as anterior cutaneous nerves, supplying the anterior abdominal wall.

As in the thorax (Figure 5.6), each nerve gives off a lateral cutaneous branch in the mid-axillary line, which pierces internal then external oblique, before supplying the anterolateral and posterolateral abdominal wall. TAP blocks (see box on *Transverse abdominal plane block*, below) performed anterior to the exit of this nerve may not anaesthetise it and therefore provide inadequate analgesia to the anterolateral/posterolateral abdominal wall.

Arteries in addition to those described above are found in the TAP, and these contribute to the vascular supply of the muscles.

Transverse abdominal plane (TAP) block

Introduction

This is a relatively new technique for providing analgesia of the anterior and lateral abdominal wall.[1,2] Local anaesthetic is introduced into the neurovascular plane between the transverse abdominal and internal oblique muscles, producing reliable analgesia of dermatomes T9/10 to L1. The block relies on spread of a large volume of local anaesthetic to the multiple small abdominal nerves within the plane. TAP blocks do not provide visceral anaesthesia or analgesia and thus cannot be used as a sole technique.

Indications

Hernia repair, open appendicectomy, Caesarean section, Pfannensteil incisions (e.g. abdominal hysterectomy) and radical prostatectomy.[3]

Analgesia from T7 has been described, but there remains debate about the reliability of the block to attain this height.[4] However, the subcostal ultrasound-guided TAP block can provide analgesia above the umbilicus.[5]

The sensory blockade may take up to 60 minutes to become maximally established, so ideally the block should be placed before the start of surgery.[6]

Technique

Landmark

Commonly performed when the patient is anaesthetised. The triangle of Petit is identified in the mid-axillary line using the surface landmarks of the posterior border of the external oblique muscle anteriorly, the anterior border of the latissimus dorsi muscle posteriorly, and the iliac crest inferiorly. A short-bevel block needle is inserted perpendicularly to the skin and two 'pops' are felt as the needle passes through the fascial extensions of the external and internal oblique muscles. The needle should now lie superficial to the transverse abdominal muscle and, after aspiration, 20 ml of local anaesthetic is injected within the plane. The concentration of local anaesthetic is varied according to the calculated maximum dose allowed. The block may then be repeated on the other side.

Ultrasound-assisted block

Posterior approach
A high-frequency probe is placed on the rectus abdominis muscle, just lateral to the midline, and the transition to the linea semilunaris and subsequent division into the three muscle layers are identified (external oblique, internal oblique, transverse abdominal). The probe is moved laterally to identify the point where the transverse abdominal muscle diminishes. The regional block needle is inserted anterior to the probe using an in-plane approach and the tip is placed in the fascial plane between the internal oblique and transverse abdominal muscles. During injection of local anaesthetic a well-defined, hypoechoic spread of local anaesthetic should be seen within the fascial plane. Formation of a patchy opacity superficial or deep to this plane suggests intramuscular injection and requires repositioning of the needle.

Subcostal approach
This is performed when analgesia of the upper abdominal wall is required. Previous experience with in-plane ultrasound techniques is required to perform this block safely.
The probe is placed under the costal margin, close to the midline, along an oblique line from the xiphoid process to the anterior superior iliac spine (ASIS). The transverse abdominal muscle can be seen deep to the rectus abdominis muscle at this level. The regional block needle is inserted in-plane medial to the probe and the tip placed between the posterior rectus sheath and the transverse abdominal muscle. Hydrodissection (aspiration followed by injection of a few millilitres of local anaesthetic) may be used to open up this potential space and advance the needle further. Moving the probe further along the line toward the ASIS allows the needle to pass under the lateral border of rectus abdominis to the origin of the external and internal oblique muscles. Up to 20 ml of local anaesthetic is carefully injected in the transverse abdominal plane along this oblique subcostal line.

Complications

- Intraperitoneal injection.
- Bowel injury.
- Hepatic injury.

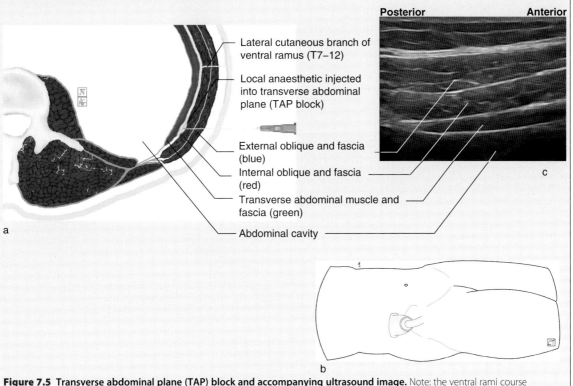

Lateral cutaneous branch of
ventral ramus (T7–12)

Local anaesthetic injected
into transverse abdominal
plane (TAP block)

External oblique and fascia
(blue)
Internal oblique and fascia
(red)
Transverse abdominal muscle and
fascia (green)
Abdominal cavity

Posterior   Anterior

a

b

c

Figure 7.5 Transverse abdominal plane (TAP) block and accompanying ultrasound image. Note: the ventral rami course anteroinferiorly through the TAP so the nerve cannot be traced back to its origin in this horizontal section.

References

1. Rafi A. Abdominal field block: a new approach via the lumbar triangle. *Anaesthesia* 2001; **56**: 1024–6.
2. McDonnell J, O'Donnell B, Tuite D, *et al.* The regional abdominal field infiltration technique computerized tomographic and anatomical identification of a novel approach to the transverse abdominis neuro-vascular fascial plane. *Anaesthesiology* 2004; **101**: A899.
3. Tran T, Ivanusic J, Hebbard P, *et al.* Determination of spread of injectate after ultrasound guided transverse abdominal plane block: a cadaveric study. *Br J Anaesth* 2009; **102**: 123–7.
4. McDonnell J, O'Donnell B, Curley G, *et al.* The analgesic efficacy of transversus abdominis plane block after abdominal surgery: a prospective randomized controlled trial. *Anesth Analg* 2007; **104**: 193–7.
5. Hebbard P, Barrington M, Vasey C. Ultrasound-guided continuous oblique subcostal transverse abdominal plane block. *Reg Anesth Pain Med* 2010; **35**: 436–41.
6. McDonnell J, O'Donnell B, Farell T, *et al.* Transverse abdominal plane block: a cadaveric and radiological evaluation. *Reg Anesth Pain Med* 2007; **32**: 399–404.

6. The inguinal region

Refers to the inferolateral portions of the abdominal wall.

The inguinal ligament is formed from the free edge of the aponeurosis of external oblique which is folded back on itself. It runs from the anterior superior iliac spine (ASIS) to the pubic tubercle.

i) The inguinal canal

A passage coursing inferomedially through the muscles of the anterior abdominal wall. Extends from the deep inguinal ring (opening in transversalis fascia), to the superficial inguinal ring (opening in external oblique).

Boundaries:

- Anterior – aponeurosis of external oblique and fibres of internal oblique.
- Posterior – transversalis fascia, merging with internal oblique medially to form the conjoint tendon.
- Roof – internal oblique and transverse abdominal muscles.
- Floor – inguinal ligament.

Contents:

- Male – ilioinguinal nerve, spermatic cord (containing the ductus deferens; the testicular, ductus deferens and cremasteric arteries; the pampiniform venous plexus; the genital branch of the genitofemoral nerve; sympathetic nerve fibres and lymphatic vessels).
- Female – ilioinguinal nerve and the round ligament of the uterus.

ii) Neural supply

- Iliohypogastric nerve (L1): sensory to skin over the upper inguinal region superior to the pubis, iliac crest and superolateral buttock; motor to internal oblique and transverse abdominal.
- Ilioinguinal nerve (L1): sensory to skin of the proximal medial thigh, anterolateral scrotum and base of penis/labium majus, mons pubis; motor to internal oblique and transverse abdominal.

The iliohypogastric and ilioinguinal nerves emerge from the lumbar plexus at the lateral border of psoas muscle (Figure 7.2). They cross quadratus lumborum and enter the transverse abdominal plane (TAP). Superomedially to the ASIS they are still located in the TAP, but as they course inferiorly they pierce internal then external oblique to provide the cutaneous sensation described.

Ilioinguinal and iliohypogastric nerve block

Introduction

This is a low-skill block that may be a useful analgesic-sparing technique.

Indications

Surgery on the lower abdominal wall and inguinal region, e.g. hernia repair. The block does not provide visceral anaesthesia, so for a hernia repair the sac containing peritoneum must be infiltrated separately. For other surgical procedures the block is used as an adjunctive technique for analgesia.

Technique

Landmark

The anterior superior iliac spine (ASIS) is marked on the skin and the needle entry point is marked 2 cm medial and 2 cm superior to it. The nerves pass consistently between the transverse abdominal and internal oblique muscles above the ASIS, but inferior to the ASIS the point at which they pass through the internal oblique is variable.[1]

A short-bevel block needle is inserted perpendicularly through the skin. A 'pop' is felt as the needle penetrates the external oblique aponeurosis, and 5 ml of local anaesthetic is injected into the plane between the external and internal oblique muscles. The needle is advanced further until a second 'pop' is felt as the needle penetrates the internal oblique aponeurosis, and a further 5 ml of local anaesthetic is injected into the plane between the internal oblique and transverse abdominal muscles (although the anatomy is consistent, the perception of 'pops' may not be, hence deposition in both planes). A further 3–5 ml of local anaesthetic may be injected subcutaneously in a fan-like distribution to block contributions from the T11–12 nerves and to overcome anatomical variation. This approach may lead to deposition of local anaesthetic more than one layer away from the nerves, but it has a success rate of about 70%.[2]

Ultrasound-assisted block

A high-frequency probe is placed on a line from the ASIS to the umbilicus. The plane between the internal oblique and transverse abdominal muscles is identified (possibly aided by scanning as for a TAP block (Figure 7.5c) and following the plane back to the ASIS). Within this plane, about 1–3 cm along the line from the ASIS to the umbilicus, the ilioinguinal and iliohypogastric nerves are identified with the ascending branch of the deep circumflex iliac artery lying between them.[3] If the nerves cannot be visualised, colour Doppler imaging can reveal the artery, and injection of local anaesthetic adjacent to the artery is likely to produce anaesthesia.

A short-bevel block needle is inserted either in-plane or out-of-plane. The tip is placed between the internal oblique and transverse abdominal muscles, and 10–15 ml of long-acting local anaesthetic is injected in aliquots with intermittent aspiration. For an in-plane approach it is probably safer to insert the needle medial to the probe, aiming towards the ASIS.

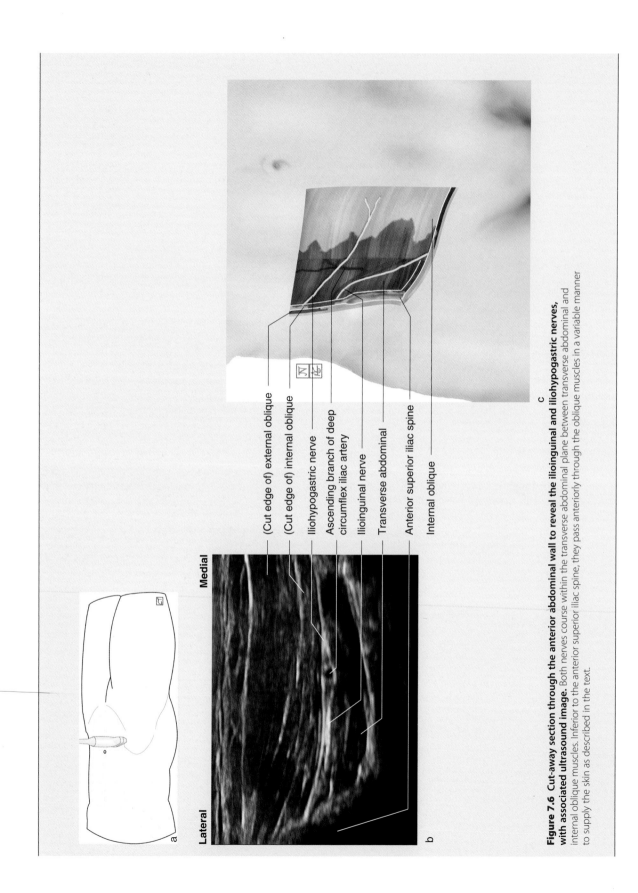

Lateral

Medial

(Cut edge of) external oblique

(Cut edge of) internal oblique

Iliohypogastric nerve

Ascending branch of deep
circumflex iliac artery

Ilioinguinal nerve

Transverse abdominal

Anterior superior iliac spine

Internal oblique

a

b

c

**Figure 7.6 Cut-away section through the anterior abdominal wall to reveal the ilioinguinal and iliohypogastric nerves,
with associated ultrasound image.** Both nerves course within the transverse abdominal plane between transverse abdominal and
internal oblique muscles. Inferior to the anterior superior iliac spine, they pass anteriorly through the oblique muscles in a variable manner
to supply the skin as described in the text.

Complications

- Perforation of bowel.
- Femoral nerve block – incidence up to 5% of landmark techniques,[4] due to spread of misplaced local anaesthetic along the fascia iliaca. Using a needle insertion point medial and inferior to the ASIS may place the needle tip in or near the inguinal ligament and increase this risk.

References

1. Al-Dabbagh A. Anatomical variations of the inguinal nerves and risks of injury in 110 hernia repairs. *Surg Radiol Anat* 2002; **24**: 102–7.
2. Weintraud M, Marhofer P, Bosonberg A, *et al.* Ilioinguinal/iliohypogastric blocks in children: where do we administer the local anaesthetic without direct visualization? *Anesth Analg* 2008; **106**: 89–93.
3. Gofield M, Christakis M. Sonographically Guided Ilioinguinal Nerve Block. *J Ultrasound Med* 2006; **25**: 1571–5.
4. Ghani KR, McMillan R, Paterson-Brown S. Transient femoral nerve palsy following ilioinguinal nerve blockade for day case inguinal hernia repair. *J R Coll Surg Edinb* 2002; **47**: 626–9.

7. The penis and scrotum

Neural supply

The penis and scrotum are supplied by the ilioinguinal and genitofemoral nerves and by branches of the pudendal nerve.

Penis:

- Ilioinguinal nerve – supplies the root or base of the penis.
- Dorsal nerve of the penis, derived from the pudendal nerve.

Scrotum:

- Anterolateral – ilioinguinal nerve (L1) and the genital branch of the genitofemoral nerve (L1, 2).
- Posterior – superficial perineal nerves derived from the pudendal nerve.

The pudendal nerve arises from the sacral plexus (S2–4) (Figure 8.14b). Having left the pelvis via the greater sciatic foramen, this nerve hooks round the ischial spine and enters the perineum through the lesser sciatic foramen. The fibres (now the dorsal nerve of the penis) pass under the symphysis pubis and enter the subpubic space beneath Buck's fascia (the deep fascia of the penis) where the suspensory ligament of the penis separates the nerves supplying left and right. This explains why a bilateral injection is needed for penile anaesthesia (Figures 7.7 and 7.8).

Penile block

Introduction

This is a low-skill block which can significantly reduce analgesic requirements.

Indications

Circumcision, release of paraphimosis or phimosis, dorsal slit of the foreskin, repair of penile lacerations and release of penile skin trapped in a zip.

Specific contraindications

- Adrenaline-containing local anaesthetic solutions (associated with penile ischaemia and necrosis).
- Suspected testicular torsion.
- Skin infection.

Pre-procedure checks

For an adult it is useful to make up a mixture of 0.5% bupivacaine 10 ml and 1% lidocaine 10 ml (both without adrenaline) up to a total volume of 30 ml with 0.9% NaCl. 10 ml may be used for the dorsal penile block, 10 ml for the ring block and the remaining volume used if an area is missed, e.g. the frenulum.

Technique

Landmark

Subpubic approach

A 27G needle is inserted under the middle of the pubic arch at the base of the penis until it touches the pubic symphysis. The depth is noted and the needle is withdrawn and redirected to pass either left or right of the midline below the pubic symphysis about a further 0.5 cm deeper until a 'pop' is felt as the needle enters Buck's fascia. After careful aspiration up to 5 ml of local anaesthetic (without adrenaline) is injected either side of the midline. Care is taken to avoid the superficial dorsal penile vein in the midline.

Subcutaneous ring block

A subcutaneous ring block is placed around the base of the penile shaft to anaesthetise the contribution from the ilioinguinal nerve. 10 ml of local anaesthetic is injected from ventral and dorsal insertion sites.

Frenulum

A dorsal penile nerve block may miss the nerves to the frenulum. 1–2 ml of local anaesthetic may be injected at the base of the ventral aspect of the penis.

Complications

- The subcutaneous ring block has a higher incidence of inadequate postoperative pain relief if used as the sole technique.

Post-procedure checks

- At least 10–15 minutes should elapse between injection and operation, to allow the block to establish.

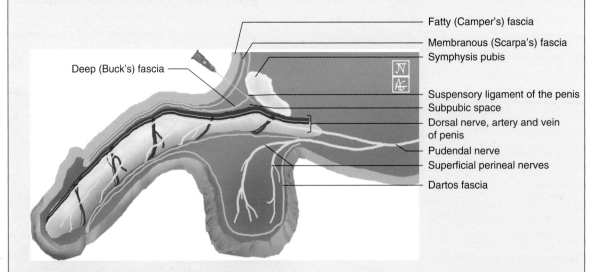

Figure 7.7 Sagittal view through the penis showing the anatomy of a penile block. The aim is deposition of local anaesthetic in the subpubic space, beneath Buck's fascia, either side of the midline. Spread of local anaesthetic in this plane ensures anaesthesia of the dorsal nerve of the penis.

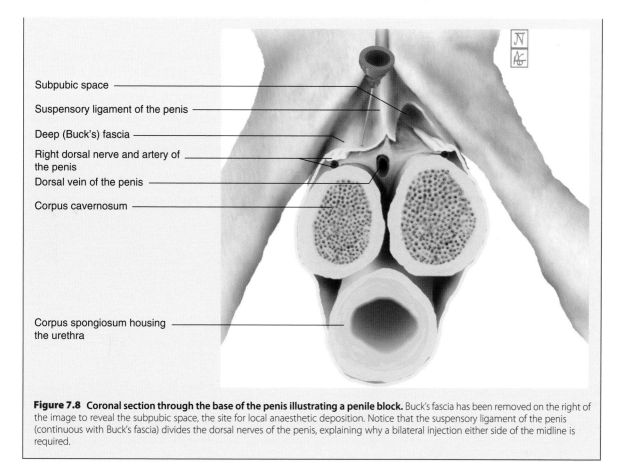

Subpubic space

Suspensory ligament of the penis

Deep (Buck's) fascia

Right dorsal nerve and artery of
the penis

Dorsal vein of the penis

Corpus cavernosum

Corpus spongiosum housing
the urethra

Figure 7.8 Coronal section through the base of the penis illustrating a penile block. Buck's fascia has been removed on the right of the image to reveal the subpubic space, the site for local anaesthetic deposition. Notice that the suspensory ligament of the penis (continuous with Buck's fascia) divides the dorsal nerves of the penis, explaining why a bilateral injection either side of the midline is required.

8. Visceral pain

Pain does not arise from all viscera per se (e.g. the liver parenchyma), but may be elicited by stretching or irritation of the associated capsule (e.g. the liver capsule).

Visceral pain has a different quality and character from somatic pain:

- Diffuse and difficult to localise.
- Commonly referred to cutaneous structures.
- Commonly associated with autonomic reflexes (e.g. nausea, sweating).
- More commonly triggered by stretch, ischaemia, inflammation and traction.

Unlike somatic pain, visceral pain reaches the central nervous system by travelling with sympathetic and/or parasympathetic nerve fibres.

i) Sympathetic visceral afferent pathway

- From the organ, the fibres travel with the associated sympathetic efferent neuron back to the sympathetic trunk.
- Here they pass via the grey rami communicantes through the corresponding paravertebral ganglion in the trunk and then via the white rami communicantes to the ventral spinal nerve (Figure 1.1).
- The fibre then diverges from the sympathetic pathway by entering the dorsal root ganglion (where the cell body is located), and from there it travels via the dorsal roots to synapse with second-order neurons in laminae I and II of the dorsal horn (Figure 2.7). This final pathway is common to both visceral and somatic fibres.

- Cutaneous afferents synapse in the same laminae, explaining why visceral pain may be referred to the dermatome corresponding to that spinal level.
- Some visceral afferents travel with sympathetic neurons, but join somatic afferent fibres before passing through the sympathetic trunk.

ii) Parasympathetic visceral afferent pathway

As with the efferent supply, the afferent pathway is craniosacral (see Chapter 1, section 7).

- Afferent fibres as far distally as two-thirds of the way along the transverse colon travel within the vagus nerve back to the solitary nucleus of the vagus (the sensory nucleus).
- Fibres distal to that point and those serving the pelvic viscera travel with the associated parasympathetic efferent neuron back to the spinal segments S2–4. Here they enter the dorsal root (cell body in dorsal root ganglion) and then the dorsal horn, where the final pathway is as described above for the sympathetic visceral afferent pathway.

The lower limb

The vessels of the lower limb

1. Arterial supply

See Figure 8.1.

Location of pulses:

- Femoral – the mid-inguinal point (midway between the anterior superior iliac spine and the pubic symphysis). Note: this is not the midpoint of the inguinal ligament.
- Popliteal – deep in the inferior part of the popliteal fossa, with the knee flexed.
- Dorsal artery of the foot (formerly dorsalis pedis) – along a path midway between the malleoli directed towards the first and second toes, lateral to extensor hallucis longus.
- Posterior tibial – between the posterior surface of the medial malleolus and the medial border of the calcaneal (Achilles) tendon with the foot relaxed.

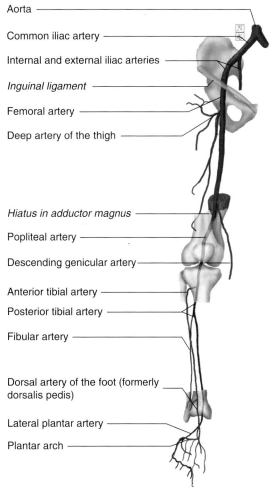

Aorta

Common iliac artery

Internal and external iliac arteries

Inguinal ligament

Femoral artery

Deep artery of the thigh

Hiatus in adductor magnus

Popliteal artery

Descending genicular artery

Anterior tibial artery

Posterior tibial artery

Fibular artery

Dorsal artery of the foot (formerly dorsalis pedis)

Lateral plantar artery

Plantar arch

Figure 8.1 Schematic illustration of the major arteries of the lower limb. Labels in italics represent the transition points from one named artery to the next.

2. Venous supply

See Figure 8.2.

Veins of the dorsal venous arch of the foot drain to superficial and deep veins of the leg, which communicate via perforating veins.

i) Superficial veins

- Great saphenous vein – passes anterior to the medial malleolus and ascends on the medial side of the leg, passing a hand's-breadth medial to the patella. It joins the femoral vein in the femoral triangle.
- Small saphenous vein – passes posterior to the lateral malleolus and ascends between the heads of gastrocnemius to empty into the popliteal vein in the popliteal fossa.

ii) Deep veins

- Accompany the arteries.

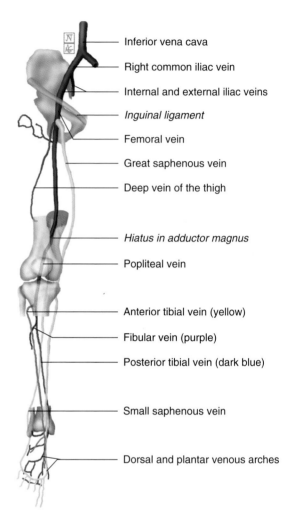

Inferior vena cava

Right common iliac vein

Internal and external iliac veins

Inguinal ligament

Femoral vein

Great saphenous vein

Deep vein of the thigh

Hiatus in adductor magnus

Popliteal vein

Anterior tibial vein (yellow)

Fibular vein (purple)

Posterior tibial vein (dark blue)

Small saphenous vein

Dorsal and plantar venous arches

Figure 8.2 Schematic illustration of the major veins of the lower limb. Labels in italics represent the transition points from one named vein to the next. The dorsal and plantar arches drain to the anterior/posterior tibial and fibular veins, which accompany their respective arteries. The superficial veins, shown in light blue, are the small saphenous vein (drains into the popliteal vein) and the great saphenous vein (drains into the femoral vein).

Fascia of the thigh

Clinically important when performing nerve blocks.

1. Superficial fascia

Found deep to the skin and blends with the dermis.

2. Deep fascia

Fascia lata – invests the limb, preventing bulging of muscles during contraction. Attaches to the inguinal ligament, iliac crests, pubis and sacrum.

Fascia iliaca – the fascial covering of iliacus and psoas major muscles, found deep to fascia lata. Runs from the lower thoracic vertebrae, down the posterior abdomen and pelvis. It attaches to the iliac crest and the anterior superior iliac spine (ASIS) then to the posterior part of the inguinal ligament before forming the posterior wall of the femoral sheath. Distally it increases in thickness (making puncture with a blunt needle quite obvious) before blending with the fascia lata.

The space between the iliacus/psoas muscles and the fascia iliaca is known as the fascia iliaca compartment. Within the compartment lie the femoral nerve, the lateral cutaneous nerve of the thigh and the obturator nerve after they exit psoas major muscle (Figure 7.2).

The sacrococcygeal plexus

The sacral plexus is formed from the ventral rami of L4, 5 and S1–4.
The coccygeal plexus is formed from the ventral rami of S4, 5 and Co1.
Supplies:
- Motor and sensory innervation to the lower limb, gluteal area and perineum.

The ventral rami of L4 and L5 unite to form the lumbosacral trunk at the medial border of psoas major. The trunk travels over the pelvic brim to join S1 within the pelvis, where the remainder of the plexus is found (Figure 8.3).

1. Terminal nerves

The main terminal nerves to leave the plexus are:
- Sciatic nerve (L4–S3) – see *Neural supply of the lower limb*, section 2, below.
- Pudendal nerve (S2–4) – see Chapter 7, section 7.

2. Collateral branches

- Motor – to the muscles of the pelvis, perineum and gluteal regions.
- Sensory – posterior cutaneous nerve of the thigh (S1, 2, 3) – supplies the buttock, medial and posterior surfaces of the thigh. Perforating cutaneous nerve (S2, 3) – supplies medial buttock.
- Visceral – pelvic splanchnic nerves (S2, 3) – pelvic viscera via inferior hypogastric and pelvic plexuses (see Chapter 1, section 8).

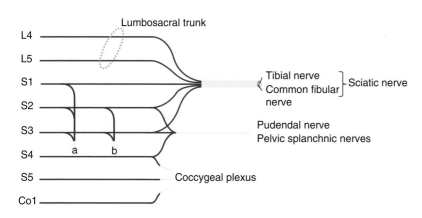

Figure 8.3 Schematic representation of the sacrococcygeal plexus. Note that the sciatic nerve is really just the tibial and common fibular nerves bound by a common connective tissue sheath.
a – posterior cutaneous nerve of the thigh;
b – perforating cutaneous nerve.

Neural supply of the lower limb

The neural supply to the lower limb is derived from the lumbar and sacrococcygeal plexuses, described in detail in Chapter 7, section 4 and under *The sacrococcygeal plexus*, above, respectively.

The following more explicitly delineates the anatomy of the major terminal nerves.

1. From the lumbar plexus

i) Iliohypogastric and ilioinguinal nerves (L1)

See Chapter 7, section 6.

The iliohypogastric nerve supplies sensation to the upper inguinal region and superolateral buttock. The ilioinguinal nerve supplies sensation to the lower/medial inguinal region and proximal thigh.

ii) Femoral nerve (L2–4)

Formed within psoas major muscle.

Enters the thigh lateral to the femoral artery. Within the femoral triangle it splits into several branches, the largest being the saphenous nerve.

- Motor – anterior thigh muscles.
- Sensory – skin of anteromedial thigh; hip and knee joints.

iii) Saphenous nerve (branch of the femoral nerve)

Leaves the femoral nerve and travels beneath sartorius muscle with the femoral artery (Figure 8.17a).

It then follows the descending genicular artery (which comes off the femoral artery prior to its descent through the adductor hiatus to become the popliteal artery) to the medial aspect of the knee.

Here it becomes very superficial and accompanies the great saphenous vein down the leg, to pass anterior to the medial malleolus, terminating in the foot.

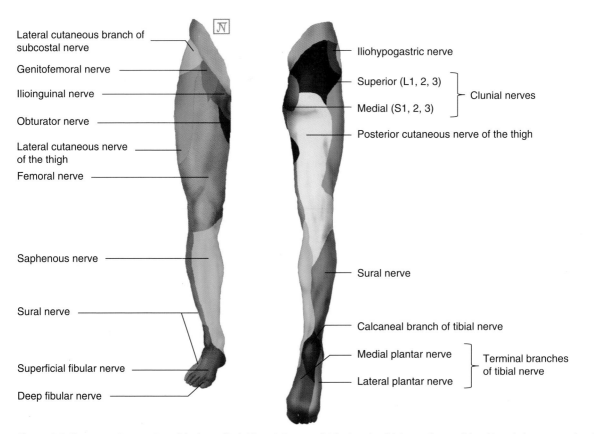

Figure 8.4 Cutaneous innervation of the lower limb. The sciatic nerve divides into the tibial nerve (→ medial and lateral plantar nerves) and the common fibular nerve (→ deep and superficial fibular nerves). A branch from the tibial and common fibular forms the sural nerve. The saphenous nerve is a continuation of the femoral nerve.

- Motor – nil.
- Sensory – skin of the medial knee, medial tibia and medial malleolus.

iv) Obturator nerve (L2–4)

Formed within psoas major muscle. Passes posterior to the common iliac vessels.
Passes through the obturator foramen.

- Motor – adductor group of muscles.
- Sensory – skin of medial (but also some anterior and posterior) thigh; hip and knee joints.

v) Lateral cutaneous nerve of the thigh (L2, 3)

Runs inferolaterally on iliacus, then posterior to the most lateral portion of the inguinal ligament.
It then lies beneath the fascia iliaca in the proximal thigh.

- Sensory – skin over the anterolateral thigh from the greater trochanter to the knee joint.

Femoral nerve block

Introduction

Femoral nerve block is relatively simple and has a low risk of complications.

Indications

Surgery on the anterior aspect of the thigh, femur or knee.[1]
Superficial surgery on the medial aspect of the lower leg, e.g. long saphenous vein stripping.

Specific contraindications

- Previous ilioinguinal surgery – renal transplant, vascular graft.
- Paraesthesia should not be sought during a femoral nerve block.

Technique

Nerve stimulator

The patient is in a supine position with the leg abducted 10–20° and slightly externally rotated. The anterior superior iliac spine (ASIS) and pubic tubercle are palpated and the position of the inguinal ligament is marked. The inguinal crease is also marked. The pulsatile femoral artery is palpated and its position marked between the two lines.

A 50 mm stimulator needle is inserted at 45° to the skin in a cephalad direction 1 cm lateral to the femoral artery and 1 cm below the inguinal ligament. Two 'pops' may be perceived as the needle penetrates the fascia lata and fascia iliaca. Patella twitches are elicited at a current between 0.2 and 0.5 mA, and 10–15 ml of local anaesthetic is injected in aliquots after negative aspiration. It is not uncommon to see twitches of the sartorius muscle rather than a patella twitch. The femoral nerve has an

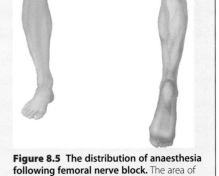

Figure 8.5 The distribution of anaesthesia following femoral nerve block. The area of anaesthesia is shown by the coloured areas, which correlate with Figure 8.4.

anteromedial component that gives rise to the branches that innervate sartorius, and a posterior component that innervates the quadriceps muscle. If sartorius twitches are elicited then the needle should be redirected more laterally and slightly deeper.

Ultrasound-assisted block

The inguinal ligament is identified as above. A high-frequency probe is placed just distal and parallel to the inguinal ligament. The pulsatile femoral artery is visualised, with the femoral nerve laterally. The fascia lata is identified superficial to the femoral vessels, and the fascia iliaca is seen immediately deep to them. The nerve appears oval or triangular and lies deep to the fascia iliaca. If the femoral artery is seen to divide (giving off the deep artery of the thigh) then the probe should be moved more proximally, as at this level the femoral nerve has divided into terminal branches that are too small to visualise.

A 50 mm stimulator needle is introduced in-plane from the lateral aspect of the leg and advanced through the two fascial layers to a point just deep to the nerve. 5–10 ml of local anaesthetic is injected. The needle may be redirected to a point superior to the nerve, but below the fascia iliaca, and a further 5–10 ml of local anaesthetic injected to hydrodissect the nerve off the fascia and encircle it. A volume of greater than 20 ml is not associated with an increased success rate.[2]

Complications

- The quadriceps muscles have reduced strength for the duration of the block,[3] which may predispose to delayed mobilisation or postoperative falls.[4]
- Vascular puncture and haematoma, which may produce nerve compression.
- Inadvertent intravascular injection.

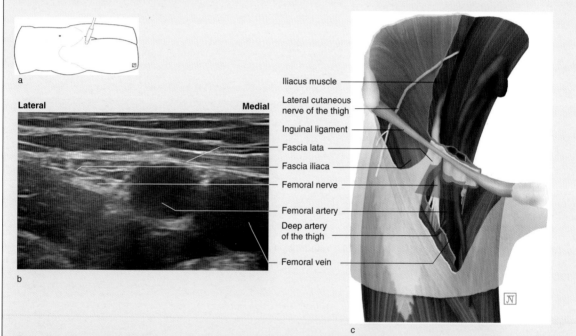

Figure 8.6 **The anatomy of a femoral nerve block and associated ultrasound image.** The beam of the ultrasound is shown by the 'pane' on the anatomical figure. Note the relative distributions of the fascias lata (latte-coloured) and iliaca (lilac-coloured) in the region of the block: local anaesthetic must be deposited beneath both fascias and above the iliacus muscle to envelop the femoral nerve. Note also that at the level of the deep artery of the thigh, the femoral nerve has divided into terminal branches (and thus represents a level too distal to perform the block).

References

1. Allen H, Liu S, Ware P, *et al.* Peripheral nerve blocks improve analgesia after total knee replacement surgery. *Anesth Analg* 1998; **87**: 93–7.
2. Seeberger M, Urwyler A. Paravascular lumbar plexus block: Block extension after femoral nerve stimulation and injection of 20 vs. 40 ml mepivacaine 10 mg/ml. *Acta Anaesthesiol Scand* 1995; **39**: 769–73.
3. McLeod G, Dale J, Robinson D, *et al.* Determination of the EC50 of levobupivacaine for femoral and sciatic perineural infusion after total knee arthroplasty. *Br J Anaesth* 2009; **102**: 528–33.
4. Ilfeld B, Duke K, Donohue M. The association between lower extremity continuous peripheral nerve blocks and patient falls after knee and hip arthroplasty. *Anesth Analg* 2010; **111**: 1552–4.

Lateral cutaneous nerve of the thigh block

Introduction

The lateral cutaneous nerve of the thigh provides sensation to the anterolateral aspect of the thigh. The nerve passes under the inguinal ligament close to the anterior superior iliac spine (ASIS) and travels laterally over the sartorius muscle (Figure 8.7).

Indications

Anaesthesia for skin grafting on the lateral aspect of the thigh.
Supplementary block for proximal femoral surgery and for tourniquet pain.

Technique

Landmark

The patient is placed in a supine position, the ASIS is palpated and the needle insertion point is marked 2 cm medial and 2 cm inferior to it. A 22G 50mm short-bevelled block needle is inserted perpendicular to the skin. A 'pop' is felt as the needle crosses the fascia lata followed by a loss of resistance to injection of local anaesthetic. 10 ml of local anaesthetic is injected in a fan-like distribution above and below the fascia lata from medial to lateral.

Ultrasound-assisted block

The patient is placed in a supine position. A high-frequency probe is placed inferior and slightly medial to the ASIS parallel to the inguinal ligament. The hypoechoic bony shadow of the ASIS and the origin of the sartorius muscle are identified. The lateral cutaneous nerve of the thigh appears as a small hypoechoic structure moving from medial to lateral over the sartorius muscle in the plane between the fascias lata and iliaca.[1] Further distally, the nerve lies lateral to sartorius and superficial to the tensor fascia lata muscle.

A 22G short-bevelled block needle is inserted in-plane or out-of-plane with the tip in the plane between the fascias lata and iliaca. Hydrodissection may enhance the visibility of the nerve. 2–3 ml of local anaesthetic is injected below the fascia lata.

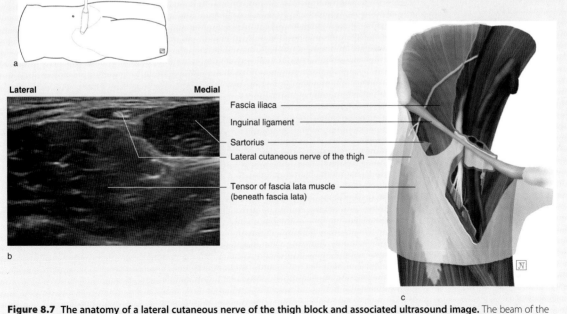

a

Lateral	Medial

Fascia iliaca
Inguinal ligament
Sartorius
Lateral cutaneous nerve of the thigh

Tensor of fascia lata muscle
(beneath fascia lata)

b

c

Figure 8.7 The anatomy of a lateral cutaneous nerve of the thigh block and associated ultrasound image. The beam of the ultrasound is shown by the 'pane' on the anatomical figure. Note that beneath the inguinal ligament the nerve lies between the fascias lata (latte-coloured) and iliaca (lilac-coloured) before passing over sartorius muscle.

Complications

- Inadvertent femoral nerve block.

References

1. Hurdle M, Weingarten T, Crisostomo R, *et al.* Ultrasound-guided blockade of the lateral femoral cutaneous nerve: technical description and review of 10 cases. *Arch Phys Med Rehabil* 2007; **88**: 1362–4.

Fascia iliaca block

Introduction

The fascia iliaca block involves injecting a large volume of local anaesthetic beneath the fascia iliaca into the potential space of the fascia iliaca compartment (see text and Figure 8.8).[1] The block is simple, quick and safe to perform and produces reliable anaesthesia of the femoral nerve and the lateral cutaneous nerve of the thigh, but does not reliably produce obturator nerve block.[2] This is in contrast to Winnie's '3–in–1 block', where injection of local anaesthetic around the femoral nerve does not dissipate proximally to a sufficient distance to the point of origin of the lateral cutaneous nerve of the thigh.

Indications

Similar to those for a femoral nerve block.
Early analgesia after acute hip or femoral fracture,[3] but ineffective for elective hip arthroplasty.[4]

Technique

Landmark

The patient is in the supine position. The anterior superior iliac spine (ASIS) and pubic tubercle are palpated and the position of the inguinal ligament is marked. The ligament is divided into three equal parts and the puncture site marked 1–2 cm caudal to the junction of the lateral third and medial two-thirds.[5] A 50 mm short-bevelled block needle is inserted perpendicular to the skin with gentle pressure exerted on the barrel of the syringe. A 'pop' is felt as the needle crosses the fascia lata, followed by a loss of resistance to injection of local anaesthetic. The needle is advanced further until another 'pop' is felt, followed by a loss of resistance to injection as the fascia iliaca is pierced. 30 ml of local anaesthetic is injected in aliquots after negative aspiration, with firm pressure applied immediately caudal to the needle to favour cephalad spread. If the characteristic 'gives' are not felt then the needle is withdrawn and a second attempt is made.

Ultrasound-assisted block

The inguinal ligament is identified by palpating the ASIS and pubic tubercle. A high-frequency probe is placed parallel to the inguinal ligament and a cross-sectional image of the femoral artery is obtained. The fascia iliaca is identified and tracked to a position 1–2 cm caudal to the junction of the lateral third and medial two-thirds of the inguinal ligament. A 50–100 mm short-bevelled block needle is inserted in-plane or out-of-plane and hydrolocation used to confirm correct needle positioning below the fascia iliaca. 30 ml of local anaesthetic is injected in aliquots after negative aspiration, and observed to flow in both a medial and lateral direction.

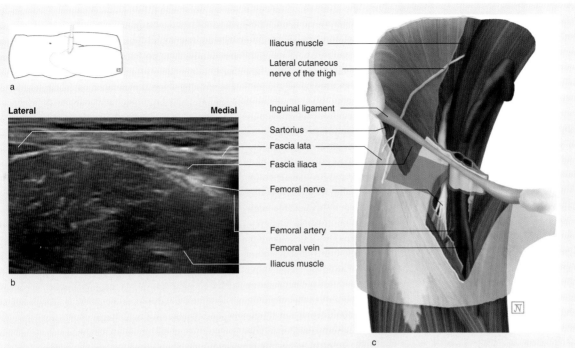

Iliacus muscle
Lateral cutaneous nerve of the thigh
Inguinal ligament
Sartorius
Fascia lata
Fascia iliaca
Femoral nerve
Femoral artery
Femoral vein
Iliacus muscle

Lateral   Medial

a

b

c

Figure 8.8 The anatomy of a fascia iliaca block and associated ultrasound image. The fascias have been cut away to reveal the anatomy beneath. The beam of the ultrasound is shown by the 'pane' on the anatomical figure. Note the relative distributions of the fascias lata (latte-coloured) and iliaca (lilac-coloured). Local anaesthetic must be deposited beneath both fascias and above iliacus muscle to spread within the fascia iliaca compartment.

Complications

- Failure due to inadequate spread of local anaesthetic.
- Inadvertent puncture of the pelvic cavity.

References

1. Dalens B, Tanguy A, Vanneuville G. Lumbar plexus blocks and lumbar plexus nerve blocks (letter to the editor). *Anesth Analg* 1989; **69**: 852–4.
2. Capdevila X, Biboulet P, Bouregba M, *et al.* Comparison of the three-in-one and fascia iliaca compartment blocks in adults: Clinical and radiological analysis. *Anesth Analg* 1998; **86**: 1039–44.
3. Foss N, Bundgaard M, Bak M, *et al.* Fascia iliaca compartment blockade for acute pain control in hip fracture patients: a randomised, placebo-controlled trial. *Anesthesiology* 2007; **106**: 773–8.
4. Shariat A, Hadzic A, Xu D, *et al.* Fascia iliaca block for analgesia after hip arthroplasty: A randomised double-blind, placebo-controlled trial. *Reg Anaes & Pain Med* 2013; **38**: 201–5.
5. Dalens B, Tanguy A. Comparison of the fascia iliaca compartment block with the 3-in-1 block in children. *Anesth Analg* 1989; **69**: 705–13.

2. From the sacrococcygeal plexus

i) Sciatic nerve (L4, 5, S1–3)

Largest nerve in the body, 2 cm wide at its origin.
Passes through the greater sciatic foramen, then deep to gluteus maximus, posterior to the acetabulum.
Passes through a point midway between the greater trochanter and the ischial tuberosity before descending in the centre of the posterior compartment of the thigh.
At the apex of the popliteal fossa, 5–12 cm proximal to the popliteal fossa crease, it divides into the common fibular (formerly common peroneal) nerve and the tibial nerve. The sciatic nerve is really just these two nerves bound together in a connective tissue sheath, with the tibial nerve medial and the common fibular nerve lateral.
The nerves may separate from one another proximal to the popliteal fossa (in 12% of people they separate prior to exiting the pelvis).

- Motor – the posterior thigh muscles and muscles distal to the knee.
- Sensory – all joints of the lower limb, all sensation distal to the knee, except for the area supplied by the saphenous nerve.

ii) Tibial nerve (branch of the sciatic nerve)

Terminal branch of the sciatic nerve.
Passes through popliteal fossa then runs inferiorly with posterior tibial vessels.
Terminates behind the medial malleolus by dividing into the medial and lateral plantar nerves, which supply the foot.

- Motor – muscles in the posterior compartment of the leg (plantar flexion and inversion).
- Sensory – knee joint.

iii) Medial and lateral plantar nerves (divisions of the tibial nerve)

- Motor – together with the deep and superficial fibular nerves, supply the intrinsic muscles of the foot.
- Sensory – see Figure 8.4.

iv) Common fibular nerve (branch of the sciatic nerve)

Terminal branch of the sciatic nerve.

Runs in the lateral popliteal fossa before winding round the neck of the fibula.
Terminates by dividing into the deep and superficial fibular nerves adjacent to the fibular neck.

- Motor – nil (motor response from deep and superficial fibular nerves).
- Sensory – knee joint.

v) Deep and superficial fibular nerves (divisions of the common fibular nerve)

The superficial fibular nerve lies in the groove between fibularis brevis and extensor digitorum longus. Distally it penetrates the fascia to become superficial and divide into branches that supply cutaneous sensation to the dorsum of the foot.
The deep fibular nerve descends on the interosseous membrane, passing anterior to the tibia at the ankle. The nerve lies with the anterior tibial artery.

- Motor – anterior muscles of the leg, and together with the medial and lateral plantar nerves supply the intrinsic muscles of the foot (producing dorsiflexion and eversion).
- Sensory – see Figure 8.4.

vi) Sural nerve (from branches of the tibial and common fibular nerves)

Arises in the popliteal fossa from a union of tibial and common fibular nerve branches.
Becomes superficial in the lower leg and accompanies the small saphenous vein to pass behind the lateral malleolus.

- Motor – nil.
- Sensory – see Figure 8.4.

vii) Posterior cutaneous nerve of the thigh (S1–3)

A direct branch of the sacrococcygeal plexus.
Runs deep to gluteus maximus and emerges from its inferior border.

- Motor – nil.
- Sensory – see Figure 8.4.

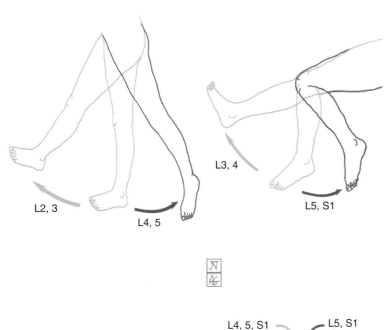

Figure 8.9 The myotomes of the lower limb.

L3, 4

L2, 3

L4, 5

L5, S1

L4, 5, S1

L5, S1

L4, 5

S1, 2

Sciatic nerve block

Introduction

A great number of approaches to the sciatic nerve have been described, some of which have questionable clinical significance. The more distal the block is performed the less likely is anaesthesia of the posterior cutaneous nerve of the thigh (a branch of the sacral plexus), which may result in discomfort if a thigh tourniquet is used. Anaesthesia in the sciatic nerve distribution may be demonstrated by weak plantarflexion of the foot. The sciatic nerve is composed of the tibial nerve and the common fibular nerve bound by a connective tissue sheath (see text).

Indications

Lower limb surgery. Most of the leg can be anaesthetised if a sciatic nerve block is combined with a femoral nerve block. Sciatic nerve block in the popliteal fossa is preferred for foot, ankle and lower leg surgery (or an ankle block for forefoot surgery).

Technique

Appropriate sedation and analgesia may be required.
10–25 ml of local anaesthetic is injected in aliquots after negative aspiration.

Landmark/nerve stimulator

A 100 mm stimulator needle is used and injection is performed when the endpoint of plantarflexion of the foot or toes (tibial nerve) is elicited at 0.2–0.5 mA. Dorsiflexion or eversion of the foot implies stimulation of the common fibular nerve, and the needle must be redirected medially.

Posterior approach (Labat)[1]

The patient is in a lateral position with the upper hip flexed and knee bent. Three lines are drawn (Figure 8.11):

- From the posterior superior iliac spine (PSIS) to the greater trochanter.
- From the sacral hiatus to the greater trochanter.
- From the midpoint of the first line, perpendicularly to the needle insertion point at the bisection of the second line (or alternatively 4 cm along the third line).

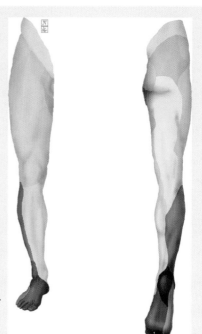

Figure 8.10 The distribution of anaesthesia following sciatic block. The area of anaesthesia is shown by the coloured areas, which correlate with Figure 8.4.

The initial twitch elicited is from the gluteal muscles. The needle is advanced deeper until the endpoint is achieved.

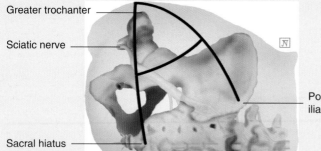

Greater trochanter

Sciatic nerve

Posterior superior iliac spine

Sacral hiatus

Figure 8.11 Schematic illustration of the posterior (Labat) approach to the sciatic nerve. The lines in this image do not appear straight as allowance has been made for the curvature of the buttock.

Parasacral approach (Mansour)[2]

The patient is positioned as for the posterior approach (above). A line is drawn between the PSIS and the ischial tuberosity. The needle insertion point is 6 cm distal to the PSIS along this line (Figure 8.14b). If bony contact is made with the needle, then the depth is noted and the needle is withdrawn and redirected slightly caudally and laterally, taking care not to advance deeper than 2 cm beyond this depth.

Inferior approach (Raj)[3]

The patient is placed supine in the lithotomy position with knee flexed and supported at 90°. The needle insertion point is halfway between the greater trochanter and the ischial tuberosity, in the groove between the hamstring and adductor muscles (Figure 8.12). The needle trajectory is perpendicular to the skin with slight medial intent.

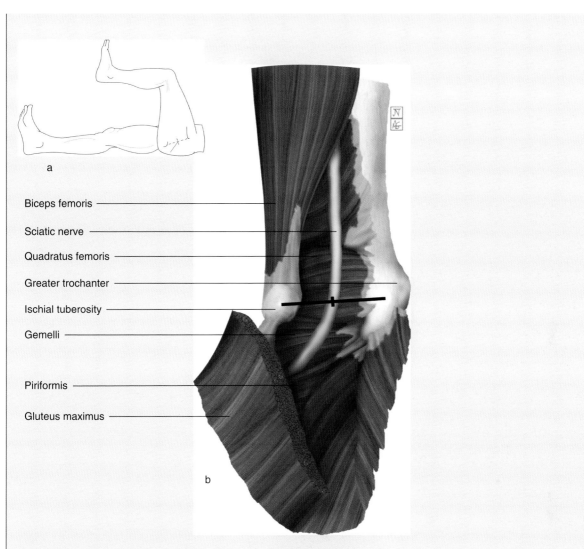

Biceps femoris

Sciatic nerve

Quadratus femoris

Greater trochanter

Ischial tuberosity

Gemelli

Piriformis

Gluteus maximus

Figure 8.12 The anatomy of the inferior (Raj) approach to the sciatic nerve.

Anterior approach (Beck)[4]
The patient is in a supine position. Three lines are drawn (Figure 8.13):
- From the anterior superior iliac spine (ASIS) to the pubic tubercle.
- From the greater trochanter, parallel to the first line.
- From the first line, at the junction of the middle and medial thirds, perpendicularly towards the needle insertion site at the bisection with the second line.

The needle trajectory is perpendicular to the skin with slight lateral intent. The ischial tuberosity may be palpated with the non-dominant hand and the needle aimed 1–2 cm lateral to this point. The femur (lesser trochanter) is often contacted, and lateral rotation of the foot may move it out of the needle path.

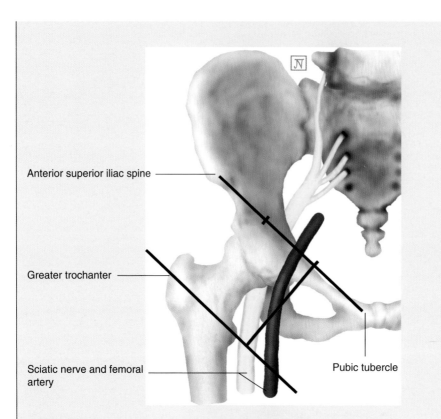

Anterior superior iliac spine

Greater trochanter

Sciatic nerve and femoral artery

Pubic tubercle

Figure 8.13 Schematic illustration of the anterior (Beck) approach to the sciatic nerve.

Ultrasound-assisted block

The sciatic nerve may be visualised from the greater sciatic notch to the popliteal fossa (Figure 8.14). The nerve is deepest at the level of the mid-thigh and easiest to see at the subgluteal or popliteal level (see box on *Sciatic nerve block in the popliteal fossa*).

Parasacral approach
The patient is in a lateral position with the upper hip flexed and knee bent. A low-frequency curved array probe is placed in a transverse plane at the level of the PSIS. The probe is moved caudally until the greater sciatic notch is located as a gap in the hyperechoic line of the ilium. The medial aspect of the probe is fixed over the sacrum and the lateral aspect of the probe is rotated more distally to visualise the posterior border of the ischium (Figure 8.14c). The sciatic nerve appears as a hyperechoic, wide, flat structure deep to gluteus maximus muscle and medial to the ischium. Colour Doppler is useful to identify the descending branch of the inferior gluteal artery, which lies immediately adjacent to the sciatic nerve.
The needle may be inserted in-plane (although the depth and angle of insertion makes needle visualisation challenging) or out-of-plane. Concurrent use of nerve stimulation is recommended. The local anaesthetic may spread to one side of the nerve, or deposit in gluteus maximus muscle, both of which require needle repositioning.

Subgluteal approach
The patient is in a semi-prone position with the limb to be blocked uppermost. The greater trochanter and ischial tuberosity are palpated and a low-frequency curved array probe is placed over the midpoint. The ischial tuberosity and femur are identified, as are gluteus maximus and quadratus femoris muscles, with the nerve lying between their fascial planes (Figure 8.14d). Concurrent use of nerve stimulation is recommended. An in-plane or out-of-plane approach may be used.

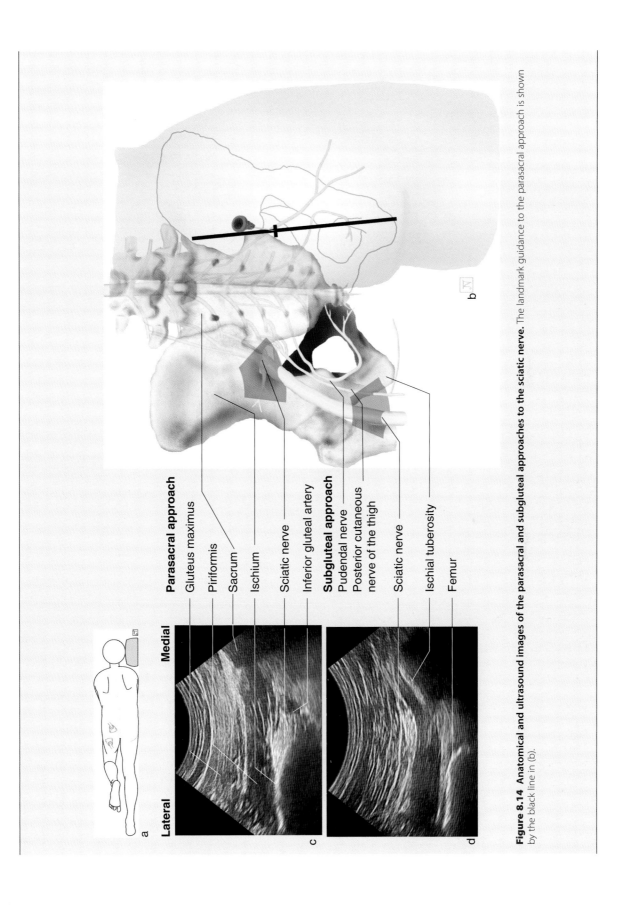

Lateral ... **Medial**

Parasacral approach

Gluteus maximus
Piriformis
Sacrum
Ischium
Sciatic nerve
Inferior gluteal artery

Subgluteal approach

Pudendal nerve
Posterior cutaneous nerve of the thigh
Sciatic nerve
Ischial tuberosity
Femur

Figure 8.14 Anatomical and ultrasound images of the parasacral and subgluteal approaches to the sciatic nerve. The landmark guidance to the parasacral approach is shown by the black line in (b).

Complications

- The sciatic nerve is uniquely susceptible to mechanical and pressure-related nerve injury.
- Care of the anaesthetised limb is paramount.
- The parasacral approach may produce transient pudendal nerve block by spread of local anaesthetic.

References

1. Labat G. *Regional Anesthesia: Its Technic and Clinical Application*. Philadelphia, PA: WB Saunders; 1924.
2. Mansour N. Reevaluating the sciatic nerve block: another landmark for consideration. *Reg Anesth Pain Med* 1993; **18**: 322–3.
3. Raj P, Parks R, Watson T *et al*. A new single-position supine approach to sciatic-femoral nerve block. *Anesth Analg* 1975; **54**: 489–93.
4. Beck G. Anterior approach to sciatic nerve block. *Anesthesiology* 1963; **24**: 222–4.

Ankle block

Introduction

The cutaneous innervation of the foot is supplied by the tibial nerve, the superficial and deep fibular nerves and the sural nerve. The saphenous nerve usually supplies skin down to the medial malleolus, but it may sometimes contribute to cutaneous innervation of the medial aspect of the foot (Figure 8.4).

Indications

Anaesthesia or analgesia for forefoot surgery.

Technique

The landmark technique is performed distally at the level of the medial and lateral malleoli with local anaesthetic infiltrated in a transverse plane through the intermalleolar line. The ultrasound technique is performed more proximally where the bony prominences have less effect on probe-to-skin contact. A 50 mm stimulator needle is used. The tibial nerve is blocked first, as it may take longest to become anaesthetised.

Landmark/nerve stimulator

Tibial nerve
The leg is externally rotated 20° and the posterior tibial artery palpated in the groove behind the medial malleolus, midway between the medial malleolus and the posterior inferior border of the calcaneum. The needle is inserted just posterior to the artery and deep to the fascia, and 5 ml of local anaesthetic is injected after negative aspiration. Nerve stimulation of the tibial nerve produces plantar flexion of the toes.

Saphenous nerve – distal approach
With the patient in the same position as above, 5 ml of local anaesthetic is infiltrated immediately above the medial malleolus and posterior to the great saphenous vein.

Saphenous nerve – proximal approach
5–10 ml of local anaesthetic is infiltrated subcutaneously at the level of the tibial tuberosity along a horizontal line from the medial tibial condyle across the medial calf.

Superficial fibular nerve
5–10 ml of local anaesthetic is infiltrated subcutaneously along the anterior intermalleolar line.

Deep fibular nerve
The deep fibular nerve enters the foot lateral to the tendon of extensor hallucis longus in close relationship to the dorsal artery of the foot. The tendon of extensor hallucis longus can be identified by asking the patient to dorsiflex the great toe, and the artery can be palpated in the groove lateral to it. The needle is inserted 2–3 cm distal to the

intermalleolar line medial to the artery until bony contact is made, then it is withdrawn slightly and 2 ml of local anaesthetic is injected below the extensor retinaculum. The needle is reinserted lateral to the artery and a further 2 ml is injected to achieve a fan-like distribution.

Sural nerve
5 ml of local anaesthetic is infiltrated subcutaneously between the lateral malleolus and the lateral border of the Achilles tendon; the nerve runs through the midpoint of these landmarks.

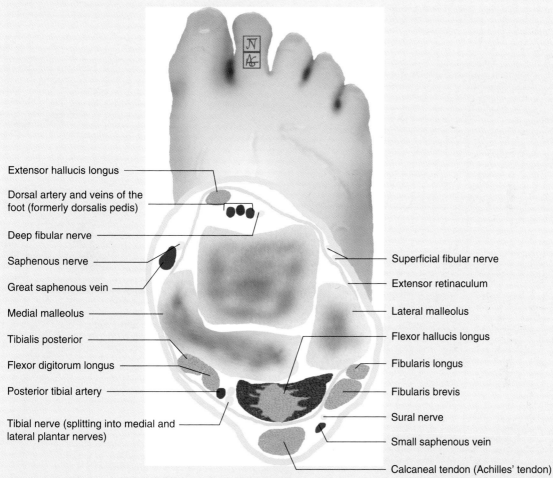

Figure 8.15 Cross-section of the ankle at the level of the malleoli, showing the structures relevant for ankle blockade.

Ultrasound-assisted block

A high-frequency probe is used. The patient is in a supine position with the ankle to be blocked either supported on a pillow or crossed over the opposite knee in the shape of a figure 4.[1]

Tibial nerve
The probe is placed above the medial malleolus, parallel to the intermalleolar line, overlying the posterior border of the tibia. The pulsatile tibial artery is identified, with the tibial nerve lying posteriorly, although occasionally it may lie anteriorly (Figure 8.16a). The block is performed in a position proximal enough to allow the needle to approach the nerve without piercing the Achilles tendon. A 50 mm stimulator needle is inserted in-plane from the posterior aspect of the leg and 5 ml of local anaesthetic is injected circumferentially around the nerve.

Saphenous nerve – distal approach

Distally, the saphenous nerve is difficult to visualise, as it is very small, but it travels down the medial aspect of the leg in close proximity to the great saphenous vein. The probe is placed lightly over the anteromedial aspect of the ankle to avoid obliterating the lumen of the vein, which is traced proximally until the nerve is easily seen. The needle is introduced in-plane from the posterior aspect of the leg and a perivascular injection is made around the great saphenous vein with 2–5 ml of local anaesthetic.

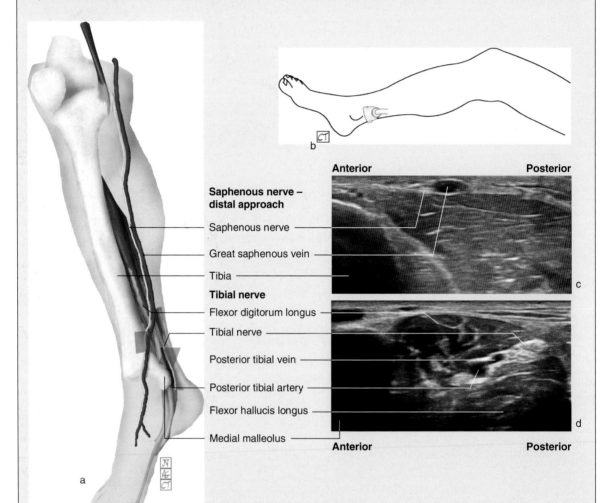

Figure 8.16 The innervation of the foot (medial view) with associated ultrasound images (ankle block). The levels at which the ultrasound images were taken are illustrated by the 'panes' on the anatomical image.

Saphenous nerve – proximal approach

It is easier to block the saphenous nerve more proximally in the leg (Figure 8.17).

The patient is in a supine position with the knee slightly flexed and the leg externally rotated to give access to the inner thigh. The probe is placed transversely at a point 10–12 cm proximal and 3–4 cm medial to the midpoint of the patella.[2] The femoral artery is identified using colour Doppler, with the sartorius muscle forming a superficial roof over it. The artery is scanned distally until it changes course to a deeper plane to become the popliteal artery. At this point the small descending genicular artery branches off and the saphenous nerve travels with it. The nerve may be visualised as a hyperechoic structure but is not always seen. The needle is inserted in-plane or out-of plane with the tip positioned in the plane between vastus medialis and sartorius. 5–10 ml of local anaesthetic is injected in aliquots after negative aspiration.

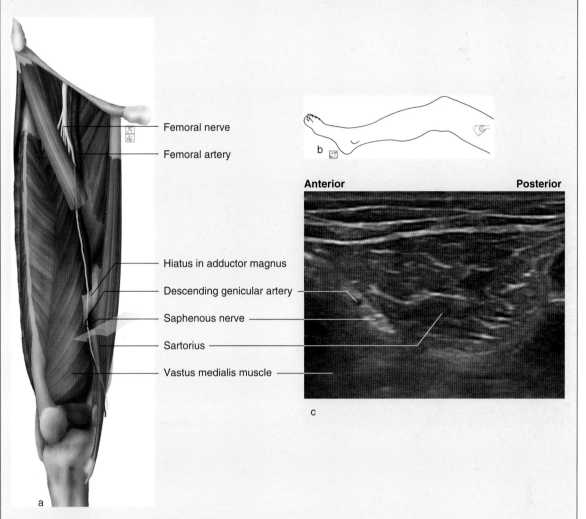

Femoral nerve

Femoral artery

Anterior   Posterior

Hiatus in adductor magnus

Descending genicular artery

Saphenous nerve

Sartorius

Vastus medialis muscle

a

b

c

Figure 8.17 The proximal anatomy of the saphenous nerve and associated ultrasound image. The beam of the ultrasound is shown by the 'pane' on the anatomical figure. The femoral artery passes through the hiatus in adductor magnus to become the popliteal artery. At this point the descending genicular artery branches off and travels with the saphenous nerve beneath sartorius (a segment of which has been removed in the anatomical image), towards the knee joint.

Superficial fibular nerve
The probe is placed just proximal to the lateral malleolus and the fibula identified with the muscle belly of extensor digitorum longus muscle (anterior) and the muscle belly of fibularis brevis (posterior) (Figure 8.18c). Scanning proximally with these two muscles centred on the screen the nerve can be seen superficial to the deep fascia, before penetrating it to lie in the groove between the two muscles. The needle is inserted in-plane from the posterior aspect of the leg, and 2 ml of local anaesthetic is injected.

Sural nerve
The probe is placed 5 cm proximal to the lateral malleolus, and the Achilles tendon (posterior) and the muscle belly of fibularis brevis (anterior) are identified (Figure 8.18d). The sural nerve lies either anterior or posterior to the small saphenous vein, which can be seen in the fascial plane between the muscles. The needle is inserted in-plane from the anterior aspect of the leg, and 2 ml of local anaesthetic is injected. Care is taken to avoid the tendon of fibularis longus.

Deep fibular nerve
The probe is placed over the anterior aspect of the tibia just proximal to the intermalleolar line. The pulsatile anterior tibial artery is visualised, often with a vein either side, and the deep fibular nerve is seen lying medial, lateral or superficial to it. 2 ml of local anaesthetic is injected above the periosteum and around the nerve.

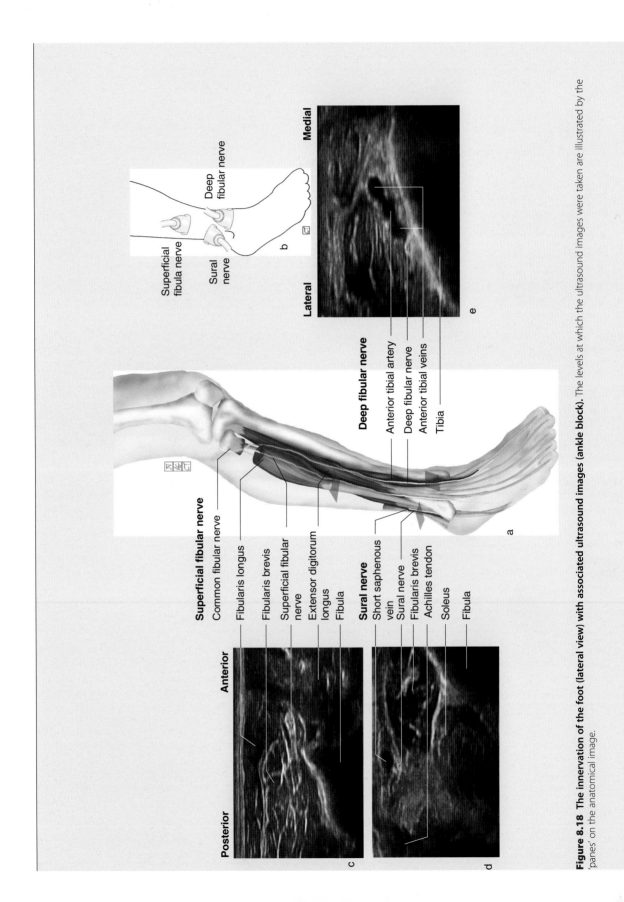

Anterior   **Posterior**

Medial

Lateral

Superficial fibular nerve

Common fibular nerve

Fibularis longus

Fibularis brevis

Superficial fibular nerve

Extensor digitorum longus

Fibula

Deep fibular nerve

Anterior tibial artery

Deep fibular nerve

Anterior tibial veins

Tibia

Sural nerve

Short saphenous vein

Sural nerve

Fibularis brevis

Achilles tendon

Soleus

Fibula

Superficial fibula nerve

Deep fibular nerve

Sural nerve

Figure 8.18 The innervation of the foot (lateral view) with associated ultrasound images (ankle block). The levels at which the ultrasound images were taken are illustrated by the 'panes' on the anatomical image.

References

1. Adriani J. *Labat's Regional Anesthesia. Techniques and Clinical Applications*, 3rd edn. Philadelphia, PA: WB Saunders; 1967, pp. 321–4.
2. Tsui B, Ozelsel T. Ultrasound-guided transartorial perifemoral artery approach for saphenous nerve block. *Reg Anesth Pain Med* 2009; **34**: 177–8.

The femoral triangle

See Figure 8.19.
A triangular fascial space in the proximal thigh.

i) Boundaries

- Floor – iliopsoas, pectineus, adductor longus (lateral to medial).
- Roof – fascia lata.
- Lateral – medial border of sartorius.
- Medial – medial border of adductor longus.
- Superior – inguinal ligament.

ii) Contents

From lateral to medial (mnemonic – N₃AVY: Nerves × 3, Artery, Vein, Y-fronts):
- Lateral cutaneous nerve of the thigh.
- Femoral branch of the genitofemoral nerve.
- Femoral nerve.
- Femoral artery.
- Femoral vein and tributaries (great saphenous and deep femoral veins).

The femoral sheath is a fascial tube which encloses the proximal parts of the femoral vessels, but does not enclose the femoral nerve. It allows vessel movement beneath the inguinal ligament during flexion of the hip.

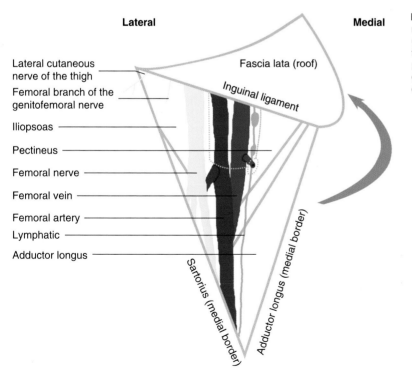

Figure 8.19 Schematic illustration of the right femoral triangle. The femoral sheath (an extension of transversalis and iliopsoas fasciae, which blends with the fascia iliaca posteriorly) is shown by the dotted line. The great saphenous vein (unlabelled) is shown draining into the femoral vein.

Femoral venous access

Introduction

The femoral vein is located medial to the femoral artery and femoral nerve (Figure 8.6c).

Indications

See box on *Internal jugular venous access*.
The femoral approach may be preferred for emergency access during cardiopulmonary resuscitation, suspected superior vena cava injury, respiratory compromise (avoids pneumothorax risk) or raised intracranial pressure.

Specific contraindications

See box on *Internal jugular venous access*.
- The femoral site is avoided if ipsilateral femoral or iliac vein thrombosis is suspected, or iliac or inferior vena cava injury.

Pre-procedure checks

See box on *Internal jugular venous access*.
A bag of fluid wrapped in a sheet placed under the ipsilateral buttock may prevent hold-up of the guidewire on the posterior wall of the vessel.

Technique

Aseptic Seldinger technique – see box on *Internal jugular venous access*.
A reverse Trendelenburg position (head-up) is employed.

Landmark

The patient is in a supine position with the leg abducted 10–20° and slightly externally rotated. The inguinal ligament is identified by palpating the anterior superior iliac spine (ASIS) and pubic tubercle. The pulsatile femoral artery is palpated at the mid-inguinal point (midway between ASIS and pubic symphysis).
The needle is inserted approximately 1 cm below the inguinal ligament and 0.5–1 cm medial to the arterial pulsation, at 45° to the skin, directed in a cephalad direction. Following vessel puncture and aspiration of venous blood, the needle is flattened out to allow easier passage of the guidewire.

Ultrasound-assisted block

The inguinal ligament is identified by palpating the ASIS and the pubic tubercle. A high-frequency probe is placed just distal and parallel to the inguinal ligament. The pulsatile femoral artery is visualised, with the femoral vein medially and the femoral nerve laterally (Figure 8.6c). The vein is usually larger, oval-shaped and compressible; the artery is pulsatile, usually smaller, more round and not easily compressible. Doppler will further help to differentiate the vessels. Difficulty in visualising the vein may be due to hypovolaemia or excessive probe pressure. If the femoral artery is seen to divide (giving off the deep artery of the thigh) then the probe should be moved more proximally, as distally the femoral vein begins to move posterior to the artery, increasing the possibility of inadvertent arterial puncture.
The vein may be approached either in-plane or out-of-plane. The out-of-plane approach allows visualisation of both vessels simultaneously, although needle appreciation, even with a relatively vertical needle intent, is limited; the in-plane approach allows good needle visualisation, but only one vessel is seen at any one time (and so this approach is recommended only for experienced practitioners). Following vessel puncture and aspiration of venous blood, the needle is flattened out to allow easier passage of the guidewire. Once the guidewire has been passed an in-plane ultrasound assessment is made to confirm correct positioning within the vein prior to dilation.

Complications

- Arterial puncture – particularly when performed during low-flow states such as circulatory arrest.
- Femoral nerve injury.
- Catheter-related bloodstream infection – the femoral site has a higher risk of infection compared with the subclavian approach, due to the proximity of the perineal area.[1]

- Haematoma and pseudoaneurysm formation.
- Thrombosis of femoral or iliac veins.
- Bowel penetration – where the needle insertion point is too cranial, or in patients with femoral hernias.
- Air embolus.
- Loss of the guidewire into the cavity.

Post-procedure checks

- Continual reassessment for catheter requirement, and routine hand washing.[1]

References

1. Pronovost P, Needham D, Berenholtz S, *et al.* An intervention to decrease catheter-related bloodstream infections in the ICU. *N Engl J Med* 2006; **355**: 2725–32.
2. Ge X, Cavallazzi R, Li C, *et al.* Central venous access sites for the prevention of venous thrombosis, stenosis and infection. *Cochrane Database Syst Rev* 2012 (3): CD004084.

The popliteal fossa

See Figure 8.20.
A diamond-shaped depression in the posterior aspect of the knee.

i) Boundaries

- Superolateral – biceps femoris.
- Superomedial – semimembranosus.
- Inferolateral and inferomedial – lateral and medial heads of gastrocnemius respectively.
- Roof – skin and fascia.
- Floor – femur, popliteal fascia, oblique popliteal ligament.

ii) Contents

- Popliteal arteries and veins.
- Small saphenous vein.
- Tibial and common fibular nerves.
- The sural nerve (although the origin of the nerve is variable).
- Posterior cutaneous nerve of the thigh.
- Popliteal lymph nodes.

Medial   **Lateral**

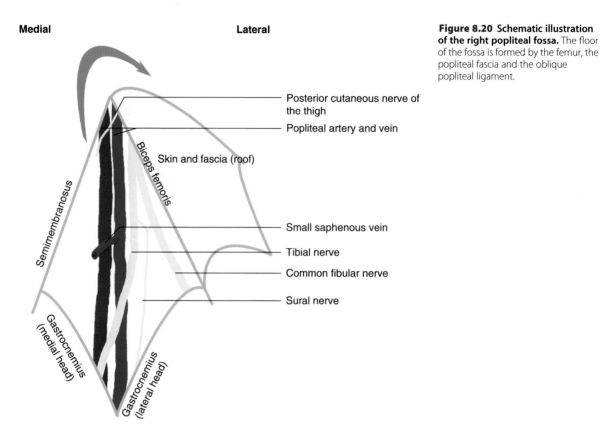

Posterior cutaneous nerve of the thigh

Popliteal artery and vein

Skin and fascia (roof)

Biceps femoris

Semimembranosus

Small saphenous vein

Tibial nerve

Common fibular nerve

Sural nerve

Gastrocnemius (medial head)

Gastrocnemius (lateral head)

Figure 8.20 Schematic illustration of the right popliteal fossa. The floor of the fossa is formed by the femur, the popliteal fascia and the oblique popliteal ligament.

Sciatic nerve block in the popliteal fossa

Introduction

Sciatic nerve block in the popliteal fossa provides anaesthesia of the calf, tibia, fibula, ankle and foot which lasts for a longer duration than an ankle block.[1] The sciatic nerve consists of two separate nerves, the tibial and common fibular nerves, which share a common epineural sheath until they diverge at about 5–12 cm proximal to the popliteal fossa crease.[2] The nerves lie superficial and lateral to the popliteal artery and vein.

Indications

Anaesthesia or analgesia for foot, ankle and lower leg surgery.

Technique

Appropriate sedation and analgesia may be required.

Nerve stimulator

Posterior approach
The patient is in a prone position with a pillow under the shin to allow easy detection of foot movement during stimulation. The needle insertion point is 7 cm above the popliteal crease and 0.5 cm lateral to the midpoint between the tendons of biceps femoris (laterally) and semitendinosus and semimembranosus (medially). A 50 mm stimulator needle is inserted, either perpendicular to the skin or at 45° aiming proximally, and advanced until either plantar flexion and inversion (tibial nerve) or dorsiflexion and eversion (common fibular nerve) of the foot are elicited. Fine movement of the needle tip from medial to lateral may be sufficient to produce alternate stimulation of each of the nerves. 20–30 ml of local anaesthetic is injected in aliquots after negative aspiration.

Lateral approach
The patient is in a supine position, with a pillow under the calf to allow easy detection of foot movement during stimulation. The posterior border of the greater trochanter is palpated and a line is drawn from this point parallel to the femur corresponding to the groove between the vastus lateralis and biceps femoris muscles. The palpating fingers are pressed into the groove about 8 cm proximal to the popliteal fossa crease and a 100 mm stimulator needle is inserted in a horizontal plane from lateral to medial. Direct stimulation of the biceps or vastus lateralis muscles may occur initially, followed by stimulation of the common fibular nerve, which lies lateral and more superficial to the tibial nerve component. 20–30 ml of local anaesthetic is injected in aliquots after negative aspiration.

Ultrasound-assisted block

The patient is in a supine or prone position. A high-frequency probe is placed in a transverse plane in the popliteal fossa, 7 cm above the popliteal crease (Figure 8.21b,c). The biceps femoris muscle is identified laterally, and semite-ndinosus and semimembranosus medially. The neurovascular structures are identified as shown in Figure 8.21. The popliteal vein is superficial to the artery but may not be seen if it collapses due to pressure from the transducer. The hyperechoic sciatic nerve is superficial to the vessels and may be better visualised if the ultrasound beam is angled caudally. The nerve is scanned proximally and distally to identify the point at which it divides into the tibial and common fibular nerves; the block is performed just proximal to this point.
A 50 mm stimulator needle is inserted, either in-plane from the lateral aspect of the thigh or out-of-plane at 45° to the skin in a cephalad direction. The needle is directed above and below the nerve from the in-plane approach or on either side of the nerve from the out-of-plane approach.
20–30 ml of local anaesthetic is injected in aliquots after negative aspiration, and it should be seen to spread circumferentially around the nerve. The nerve is often easier to visualise after local anaesthetic has been injected. Plantarflexion and dorsiflexion of the foot may produce movement of the tibial and fibular components, which assists with visualisation (the 'seesaw' sign).[3]

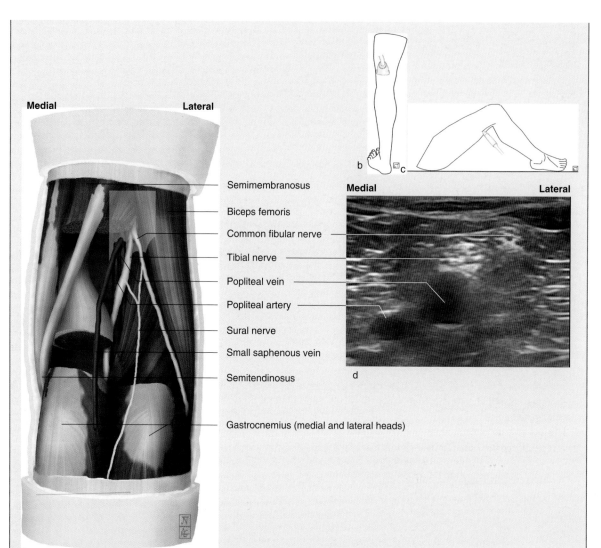

Figure 8.21 The popliteal fossa with associated ultrasound image. Note that the sciatic nerve is really just the common fibular and tibial nerves bound together by a fascial sheath. The nerves are seen to separate in this image and are distinguished on ultrasound. Probe positions are with the patient (b) prone and (c) supine.

Complications

- The sciatic nerve is uniquely susceptible to mechanical and pressure-related nerve injury.
- Vascular puncture.

References

1. McCleod DH, Wong DH, Claridge RJ, Merrick PM. Lateral popliteal sciatic nerve block compared with subcutaneous infiltration for analgesia following foot surgery. *Can J Anaesth* 1994; **41**: 673–6.
2. Vloka JD, Hadzić A, April E, Thys DM. The division of the sciatic nerve in the popliteal fossa and its possible implications in the popliteal nerve blockade. *Anesth Analg* 2001; **92**: 215–17.
3. Schafhalter-Zoppoth I, Younger SJ, Collins AB, Gray AT. The 'seesaw' sign: improved sonographic identification of the sciatic nerve. *Anesthesiology* 2004; **101**: 808–9.

The fetus

1. Requirements of the fetal circulation

Efficient delivery of the most oxygenated blood to the vital organs of the fetus (heart and brain).

Bypass of the fluid-filled lungs (physiologically redundant).

Ability to rapidly convert to adult circulation at birth.

2. Anatomical flow

- Oxygenated blood (SaO_2 80%, pO_2 4 kPa) leaves the placenta via a single umbilical vein and enters the inferior vena cava (IVC).
- 60% of IVC flow bypasses the liver via the ductus venosus, reuniting with the remaining 40% prior to entering the right atrium.
- The Eustachian valve (at the junction of the IVC and right atrium) maintains some separation of the ductus venosus blood (SaO_2 80%) and the blood from the lower half of the body (SvO_2 25–40%). The crista terminalis preferentially directs the former across the foramen ovale (an interatrial opening) into the left atrium and the latter into the right ventricle across the tricuspid valve. The crista terminalis originates at the junction of the right atrium and superior vena cava (SVC) and runs longitudinally toward the IVC.
- The blood in the left atrium (SaO_2 65%) passes to the left ventricle, and by ejection into the aorta ensures that the most oxygenated blood perfuses the heart and brain.
- Blood from the SVC (SvO_2 40%) is preferentially directed by the crista terminalis into the right ventricle, where it mixes with blood from the lower half of the body (final SvO_2 50%). Because of the high pulmonary vascular resistance, only 12% of right ventricular blood enters the pulmonary circulation; the remaining 88% enters the ductus arteriosus, which joins the descending aorta distal to the left subclavian artery (SaO_2 60%, pO_2 2.7 kPa). In this way, the least oxygenated blood supplies organs other than the heart and brain.
- Two umbilical arteries (one from each internal iliac artery) supply the placenta with 60% of the cardiac output.

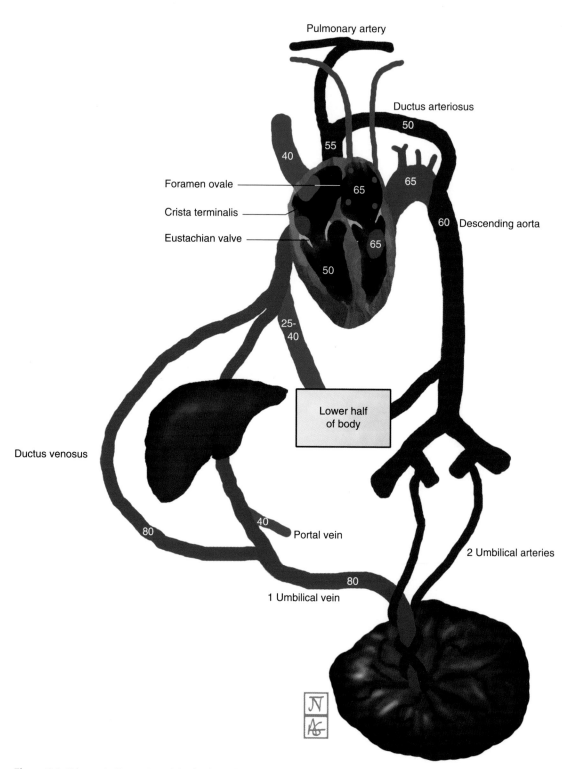

Figure 9.1 Schematic illustration of the fetal circulation. Numbers refer to the saturation of blood with oxygen at that point in the circulation.

References

Academy of Medical Royal Colleges. *A Code of Practice for the Diagnosis and Confirmation of Death*. London: The Academy; 2008.

Bergman RA, Afifi AK, Miyauchi R. Circle of Willis, usual and unusual. In Illustrated Encyclopedia of Human Anatomic Variation: Opus II: Cardiovascular System. http://www.anatomyatlases.org/AnatomicVariants/Cardiovascular/Images0200/0291.shtml (accessed November 2013).

Campbell W, DeJong R. Motor strength and power. In *DeJong's Neurologic Examination*. Philadelphia, PA: Lippincott Williams & Wilkins; 2005, pp. 387–9.

Greathouse DG, Halle JS, Dalley AF. Terminologia anatomica: revised anatomical terminology. *J Orthop Sports Phys Ther* 2004; **34**: 363–7.

Loeser J, Butler S, Chapman C, Turk D (eds.). *Bonica's Management of Pain*. Philadelphia, PA: Lippincott Williams & Wilkins; 2001, pp. 1893–952.

Moore KL, Dalley AF. Deep gluteal nerves. In *Clinically Orientated Anatomy*, 4th edn. Philadelphia, PA: Lippincott Williams & Wilkins; 1999, pp. 558–9.

Murphy PJ. The fetal circulation. *Contin Educ Anaesth Crit Care Pain* 2005; **5**: 107–12.

Power I, Kam P. The central and peripheral nervous systems. In *Principles of Physiology for the Anaesthetist*, 2nd edn. London: Hodder; 2008: pp. 46–65.

Vloka JD, Hadzić A, April E, Thys DM. The division of the sciatic nerve in the popliteal fossa and its possible implications in the popliteal nerve blockade. *Anesth Analg* 2001; **92**: 215–17.

Index